50th
Project AIR FORCE
1946–1996

RUSSIA'S AIR POWER AT THE CROSSROADS

BENJAMIN S. LAMBETH

Prepared for the
United States Air Force

RAND

PREFACE

This report is an assessment of trends and prospects in Russian military aviation. It is based in large measure on the extensive reportage on air power and other military matters that has pervaded the Russian defense literature since the onset of glasnost in 1986. It also benefits from limited first-hand contacts between the author and senior Russian Air Force and aviation industry leaders.

The roots of this undertaking go back to a Project AIR FORCE study initiated at RAND in 1987 for the Office of the Deputy Chief of Staff for Plans and Operations, Headquarters United States Air Force, aimed at providing a comprehensive look at how tactical aviation fitted into broader Soviet concepts for theater warfare in Europe. That project, entitled "Soviet Tactical Air Power in Strategic Perspective," sought to examine the USSR's fighter force in terms of its historical origins, organizational development, training and tactics, operational style, mission tasking, and role in overall theater campaign strategy.

Soon after work began, most of the original questions posed, along with the geostrategic setting of Soviet force planning, underwent a fundamental change as a result of then-President Mikhail Gorbachev's doctrinal innovations and the ensuing end of the cold war. At the same time, Gorbachev's domestic reforms created new issues of interest with regard to Soviet military aviation. They also opened up some unprecedented sources of insight into the inner workings of the Soviet defense establishment. As a result, the sponsoring office, Project CHECKMATE, agreed to put the study on hold so that these developments might be taken into proper account.

In the ensuing period, the author established contact with the chief test pilot of the Mikoyan Design Bureau at the 1988 Farnborough Air Show. As a result of that contact, in December 1989 at Kubinka Air Base near Moscow, the author became the first American citizen to fly the Soviet MiG-29 fighter and the first Westerner invited to fly a combat aircraft of any type inside Soviet airspace since the end of World War II.[1]

That experience opened further doors into the Soviet Air Force and aviation industry. During the time of the abortive August 1991 Soviet coup attempt, the author was in Moscow for the annual Aviation Day events on invitation from the head of the Central Aerohydrodynamics Institute. The following December, he had an opportunity to meet at length with the commander in chief of the Soviet Air Force, then–Colonel General Petr Deinekin, at his Moscow headquarters on the day before the heads of the Soviet republics announced the dissolution of the USSR.

The warmth and candor of that session with General Deinekin encouraged the author to seek Project AIR FORCE support for a *de novo* study of developments in Russian military aviation in the post-Soviet period that might draw on Russian Air Force cooperation. The declared intent was to provide Western air power specialists with a first-hand assessment that transcended dry facts to impart a sense of the Russian Air Force as an evolving institution.

Ultimately, the hoped-for support from the Russian Air Force proved too ambitious a goal in the still turbulent and unsteady process of post-Soviet Russian reform. Nevertheless, as an experiment in post-cold war professional outreach, the effort was not a complete failure. The author met three times more with General Deinekin during subsequent trips to Moscow to describe his research plan in detail. He also had several conversations with the head of the Russian Air Force's Central Research Institute, Major General Vasily Aleksandrov, in an attempt to build a bridge between RAND and its closest analogue in the Russian Air Force. Finally, he was able to gain additional

---

[1]A full account of that relationship and a technical overview of the MiG-29, including a detailed flight report, are presented in Benjamin S. Lambeth, *From Farnborough to Kubinka: An American MiG-29 Experience*, Santa Monica, California, RAND R-4000-RC, 1991.

insights into the operational side of Russian fighter aviation through an invitational Su-27 flight with the Gromov Flight Research Institute and two front-seat advanced handling flights in a MiG-21 and MiG-23 with the Mikoyan Design Bureau, all at the Ramenskoye Flight Test Center.[2]

The present report was prepared in the Strategy and Doctrine Program of RAND's Project AIR FORCE. It seeks to impart to the USAF leadership a richer portrait of the setting in which their Russian counterparts operate, for the purpose of better informing U.S. participation in the continuing, if now increasingly halting, exchange relationship between the two services. It should be of interest to USAF officers and other members of the defense community concerned with air operations and training, force development, comparative military capability and policy assessment, and Russian-American security relations. The bulk of the study embodies work done in support of a Project AIR FORCE research effort entitled "The World's Air Forces."[3] Chapters Four through Eight were sponsored by a subsequent project entitled "Sources of Conflict and Their Implications for Air Force Operations." The latter has been tasked to consider alternative scenarios of the global conflict arena to the year 2020, including Russia's role in it, aimed at identifying and bounding those external factors most likely to challenge the USAF through the first decade of the next century. Both projects were sponsored by the Director of Plans, Office of the Deputy Chief of Staff for Plans and Operations, Headquarters United States Air Force.

PROJECT AIR FORCE

Project AIR FORCE, a division of RAND, is the Air Force federally funded research and development center (FFRDC) for studies and analyses. It provides the Air Force with independent analyses of policy alternatives affecting the development, employment, combat readiness, and support of current and future aerospace forces. Research is being performed in three programs: Strategy and

---

[2]For a report on the latter, see Benjamin S. Lambeth, "Flying the Flogger," *Flight International*, February 23–March 1, 1994, pp. 38–41.

[3]An overview of research conducted in that larger study is presented in Christopher J. Bowie et al., *Trends in the Global Balance of Air Power*, RAND MR-478/1-AF, 1995.

Doctrine; Force Modernization and Employment; and Resource Management and System Acquisition.

In 1996, Project AIR FORCE is celebrating 50 years of service to the United States Air Force. Project AIR FORCE began in March 1946 as Project RAND at Douglas Aircraft Company, under contract to the Army Air Forces. Two years later, the project became the foundation of a new, private nonprofit institution to improve public policy through research and analysis for the public welfare and security of the United States—what is known today as RAND.

CONTENTS

FIGURES

SUMMARY

Military aviation has enjoyed a long and rich tradition in Russia, predating and thus far surviving the 74-year intercession of Soviet communism. Yet despite this rich background, until recently most Westerners have been able to follow developments in Soviet air power only from a distance, because of the USSR's obsession with secrecy. The Soviet Air Force (*Voenno-vozdushniye sily*, or VVS) was a denied area, a central component of a military threat, and an object of special Western intelligence interest throughout the cold war. Other observers had to view it darkly in an effort to understand what was going on beneath the often tantalizing, but rarely satisfying, appearances on the surface.

Today, with the cold war over and Russia embarked on a fitful quest for reform, monitoring trends in Russian military aviation has become considerably easier. Naturally, there remain limits on such inquiry. Nevertheless, it has become increasingly possible to study Russian air power much as one would study military aviation in any country.

This newly opened door has revealed a Russian Air Force in the grips of a painful metamorphosis. It is unmistakably committed to reform, yet it remains unsure of its future as it strives to embrace the 21st century as a renewed institution. For one thing, the end of the cold war and the demise of the Warsaw Pact left the VVS with no clear mission beyond homeland defense. The "Warsaw Pact Air Operation," for which Soviet pilots and commanders had purportedly planned and trained for years, was rendered moot overnight.

Beyond that, the freedom of expression made possible by glasnost prompted an unprecedented venting of complaints throughout the armed forces. As a result, the VVS found itself besieged by a multitude of pressures from below. Among the issues with which its leadership now had to contend were:

- Declining aircrew morale and retention
- An eroding quality and number of applicants for pilot training
- Severe housing shortages and an appalling quality of life for aircrews and their families
- A growing acknowledgment of deficiencies in tactical training
- Mindless administrative tyranny at the operational level
- Problems of dishonesty and lack of integrity within flying units
- Increasing concerns about the quality and reliability of the equipment provided by the aviation industry
- A mounting recognition among Russian pilots of the substantial inequalities between their own and Western equipment and training.

So what, one might ask? Why should anyone care any longer about an air force that not only has ceased being a threat to Western security, but indeed that finds itself operating in virtually a survival mode today? For one thing, the frank admissions of VVS pilots up and down the rank hierarchy since the USSR's collapse have given the West an unprecedented opportunity to update and, where necessary, to correct its past impressions of the VVS.

Beyond that, better knowledge of where the air arm of the former USSR stands today can shed useful light on the future course it may take, once the current post-Soviet reform effort reaches a more even keel. Whatever difficulties the VVS may be experiencing today, there is little doubt that Russia will eventually emerge from the collapse of Soviet communism as a strong nation. There is also little doubt that the VVS, for better or worse, will constitute an important part of Russia's military capability.

The latter point takes on added weight in light of the similarity between many of the problems now faced by the VVS and those that have come to afflict its most advanced Western counterparts. Among the principal concerns confronting the VVS today are:

- Harnessing new technology to mission requirements at an affordable cost
- Enlisting and retaining high-quality people
- Keeping the defense industry accountable to the expectations of VVS planners
- Sustaining the morale and motivation of VVS personnel at a time of major budget and force reductions
- Ensuring that the VVS develops the organizational adaptability it will need to survive as a healthy institution in the 21st century.

These problems are not so very different from those confronting most Western air forces today, the USAF included. They are aggravated severely in the case of the VVS, however, by Russia's continuing economic crisis. It should thus be instructive for USAF planners to observe how their Russian counterparts are grappling with similar issues—and in a far more stressful situation than anything we in the West have had to confront, at least yet.

CURRENT PROBLEMS AND PRIORITIES

The VVS's commander in chief, General Petr S. Deinekin, faces an array of challenges more imposing than those confronting any other air force leader in the world today. Principal among them is a complete inversion of the priorities that typically concern a peacetime military aviation establishment. Matters like force modernization, training and tactics, and similar mission-related preoccupations have taken a back seat to the more pressing demands of simply housing and caring for deprived personnel at a time when the VVS is struggling to extract itself from the wreckage of Soviet communism.

Among other things, General Deinekin is wrestling with:

- A severely curtailed procurement and operations budget
- A fuel shortage of crisis proportions

- A bloated pilot-to-aircraft ratio, further aggravating the insufficiency of available flying hours for Russian aircrews
- Widespread maintenance problems caused by a dearth of spare parts and the failure of the conscription system
- A heightened accident rate as a result of these influences
- A sharp drop in the prestige and respectability of military service, with disturbing implications for future officer recruitment.

The pilot-to-aircraft ratio in the VVS has more than doubled as a result of the withdrawal and deactivation of line units from Eastern Europe. In some regiments, the crew ratio is as great as five to one. Coupled with skyrocketing fuel costs and a marked decline in fuel allocations, this has caused a sharp reduction in available flying hours. Especially hard hit have been fighter pilots, who are now averaging only 25 flying hours a year—just enough to be dangerous to themselves.

Maintenance is also suffering, with enlisted manning down to the 50-percent level in many units and spare parts in increasingly short supply.

Finally, the decline in the former appeal of service life has undermined officer recruitment and raised hard questions about how the VVS will secure its successor generation. During the Soviet era, VVS undergraduate flight schools typically got eight or more applicants for each vacancy. Today, the application rate is scarcely better than one for one. As a result, there is no competition for available slots.

ORGANIZATION AND FORCE DEVELOPMENT

The VVS has undergone a massive drawdown over the past four years, consolidating its functions and developing new concepts of operations more appropriate to Russia's post-cold war security situation. General Deinekin has reorganized his air force around four major commands: Long-Range Aviation, Military Transport Aviation, a new Frontal Aviation Command, and a Reserve and Training Command. Previously, Russia's tactical air arm was subordinated to the military districts under army control. Now it reports directly to VVS headquarters.

It remains too early to say what the VVS will look like once the initial round of post-Soviet retrenchment is completed. Appropriations for procurement have fallen precipitously since the USSR's breakup. During the early 1980s, the Soviet air and air defense forces together acquired an average of 400–450 fighters a year. By contrast, only 23 new aircraft were purchased altogether in 1993–1994, and none were budgeted for 1996.

Frontal Aviation has shrunk from a high of over 5000 combat aircraft in 1989 to less than half that number today. Of these, around a third are fourth-generation MiG-29s and Su-27s. The remainder are older aircraft slated to be retired before the end of this decade. If current budget trends continue, Frontal Aviation's holdings, by the VVS's own estimate, will decline from 2280 to 1670 combat aircraft in 1997; to 1440 in 2000; to 1330 in 2005; to 1140 in 2010; and to 870 by 2015.

Long-Range Aviation (LRA) has shed much, though not all, of its intercontinental nuclear attack role and replaced it with the new mission of providing strategic reach in support of Russia's new regional power projection needs. It too has experienced a significant drawdown since the late 1980s as a result of arms reduction agreements and the USSR's collapse. LRA currently maintains an inventory of some 400 aircraft. It lost many of its most modern bombers to the newly independent states. The sole operational unit of Tu-160s was based in Ukraine. Upon the USSR's dissolution, these fell into Ukrainian possession and triggered a long and bitter dispute over their return to Russia as a part of the Commonwealth of Independent States' strategic forces. Only three serviceable Tu-160s were retained in Russia following the USSR's collapse. Subsequent negotiations between Moscow and Kiev, however, have resulted in the return of ten of these to the VVS at a reasonably equitable price.

The most painful loss for the VVS in the wake of the USSR's breakup was registered in Military Transport Aviation (VTA). A large portion of its Il-76 jet transports (200 out of the 450 possessed by the USSR) were based in Ukraine. This loss was especially acute in light of Russia's newly emergent regional peacekeeping challenges and the mobility requirements of Russia's new military doctrine.

Russia's Air Defense Force has experienced a rate of decline much like that of the other combat air arms since the late 1980s. From a high of some 2300 interceptors on the eve of the USSR's collapse, it is down to less than half that number today. Force modernization plans appear limited to developing and producing an advanced MiG-31, if and when procurement authorizations permit.

UNDERGRADUATE PILOT TRAINING

Throughout most of the cold war, the VVS operated a specialized undergraduate pilot training (UPT) program consisting of a dozen Higher Military Aviation Schools for Pilots. Each school trained cadets for conversion to a specific aircraft category, with curricula tailored toward fighters, ground attack aircraft, strategic bombers, or transports, depending on the school. Application meant signing up for a 25-year service commitment in the event of successful completion of the program, with service time beginning with the year of enrollment. The curriculum covered four years and included training in officership, basic college-level science and engineering, and entry-level aeronautical skills. Graduates were commissioned with the rank of lieutenant, awarded a degree in engineering, and given an aeronautical rating of basic pilot.

Through the mid-1980s, the VVS provided its prospective fighter pilots with initial qualification training on their assigned aircraft during the final year of UPT. Cadets were cleared to fly the MiG-21 or MiG-23 only within a narrow band of operations, to include clear-weather takeoffs and landings, formation flight, and basic ground attack and air combat maneuvering. They would then report directly to their assigned unit for a lengthy upgrading to mission-ready status. This approach meant that every line fighter regiment commander was responsible not only for his unit's combat capability, but also for a full conversion training program that typically worked at cross-purposes with his readiness needs.

A pronounced weakness of the Soviet approach to pilot selection was its provision that allowed the aspirant himself to select the aircraft category for which he would compete. Such an approach may have ensured that only the most capable cadets accepted into any UPT school would complete the program and earn wings. However, because it did nothing to stream the right *kinds* of candidates to those schools where they would be best suited by temperament and abilities, it had no way of guaranteeing that the best people in the country for fighter training, for example, would be properly identified and vectored toward such training.

By 1988, the VVS had abandoned initial operational conversion in the UPT schools and shifted to instructing on a single aircraft type (either the L-29 or L-39), with graduates proceeding to the equivalent of a USAF replacement training unit for combat crew training on their assigned aircraft. The purpose of this changed approach was to relieve line unit commanders of much of their former conversion training burden. Since the collapse of the USSR, VVS undergraduate pilot training has undergone even further change. The new system approved by General Deinekin entails initial flight orientation in secondary boarding schools, followed by screening and selection for a *five*-year UPT program, with flight training solely on the L-39. Now cadets receive three years of classroom academic instruction. Only afterward is it determined whether they will continue on to the flying phase, which is now compressed into two years rather than spread out over three as before. Of the twelve original UPT schools, the VVS now operates only four.

CONTINUATION TRAINING IN LINE FIGHTER UNITS

VVS fighter regiments operate very differently from their USAF counterparts. Although they are no less overburdened by paperwork and reporting responsibilities, they are far less consistent about honoring published rules in day-to-day practice. VVS headquarters promulgates explicit and detailed standardization regulations. Yet their observance is heavily dependent on the personal inclinations of the regimental commander.

Flight activity at the squadron level does not take place daily over a five-day week as in most Western air forces, but rather every other day, with alternating maintenance down days over a six-day week. At the height of the cold war, when operations and support funding was relatively unconstrained, the operating tempo of fighter units on flying days was considerably higher than that of a USAF wing. Pilots would typically fly three or four sorties of about 30 minutes each, with as many as nine maintenance turns rather than just two or three.

VVS fighter operations also diverge substantially from the familiar practices of the USAF and most NATO air forces. The biggest differences are an unbending top-down imposition of stifling rules and restrictions for virtually every condition of flight, a continued denial of much latitude for individual initiative on the part of pilots, and a dominant role played by the ground command post and flight director in overseeing both conversion and continuation training.

The flight supervision group is the approximate VVS counterpart to a USAF fighter wing command post. However, its role is far more intrusive and overbearing. It features a glassed-in control center with a clear view of the ramp, taxiway, and runway at each base. It is manned during flight operations by a flight director (generally a senior pilot with the rank of major or higher), a radar controller, and a regimental-level supervisor. It combines the USAF functions of supervisor of flying, ground-controlled intercept, control tower, and runway supervisory officer. Most important, it makes many decisions for the airborne pilot that are typically made in the cockpit in Western practice.

Until the last days of the Soviet Air Force, paperwork overwhelmed daily life at the unit level. The resultant plethora of red tape and the petty micromanagement that all too often accompanied it reflected an innate distrust by VVS leaders in the professionalism and competence of their subordinates. We know now from the first-hand testimony of Russian pilots that the profusion of overlapping rules imposed on commanders was such that, were a commander to follow each restriction to the letter, he would be hard put to generate a single sortie and remain legal. This obliged commanders to maintain a double standard and live a lie on a daily basis, paying lip service to the rules in their reporting to higher headquarters, while doing whatever they felt they had to do to meet their actual training needs. Despite surface appearances, this made for little standardization among units as a practical matter.

Today, with the Soviet system repudiated and a new horizon ahead, the VVS stands on the threshold of *potentially* the most radical departure from its old ways since the earliest days of the Soviet state. Throughout VVS history, the main problem was the Soviet system, not the individual pilot or his equipment. The Soviet pilot was selected by exacting criteria, and he represented the best talent for his calling that Soviet society had to offer. Soviet aircraft and air-to-air missiles have always been respectable from a technical standpoint. The reason the VVS has long had such trouble getting the most out of these assets is that the Soviet pilot was inevitably a product of his training environment. Naturally, his techniques and skills were heavily conditioned—and circumscribed—by the inhibiting influence of a uniquely "Soviet" operational culture.

Since the USSR's collapse, the VVS has been freed of the Soviet organizational chokehold that limited its capacity to innovate. In principle, it is now at liberty to cast aside its old ways and develop a new operational repertoire aimed at extracting the fullest leverage from its highly capable equipment. Because old habits die hard, however, the VVS will most likely have great difficulty freeing itself from the top-down rigidity in both operations and thought that the communist system, for years, imposed on pilots and commanders who knew better but were obliged to pretend otherwise. Worse yet, with a shoestring budget that forces unit commanders to great exertions simply to maintain their pilots' basic aircraft handling proficiency, it is unlikely that the VVS will be able to pursue the sort of structured training—from the simple to the complex—that would be required, at a minimum, to bring Russia's pilots up to accepted Western standards.

RUSSIA'S AIR WAR IN CHECHNYA

The VVS received its first combat test during Russia's opening of operations against the secessionist republic of Chechnya. Its mission entailed backstopping Russian ground troops in putting down a local rebellion. The operational challenge was more like what NATO recently faced in Bosnia than what the allied coalition had to contend with in Operation Desert Storm. Nevertheless, Chechnya provided a relatively low-risk laboratory for testing the VVS's operational capabilities under live-fire conditions.

Russia's airlift performed commendably. VVS combat aircraft also did well in unopposed ground attacks against unsheltered Chechen aircraft. However, as the initial ground campaign unfolded and stresses mounted because of weather complications and the demand for high-accuracy bombing in the face of effective low-altitude defenses, degradations in performance were displayed repeatedly by VVS aircrews, attesting to training shortcuts caused by a lack of money.

Early in the war, reported bombing inaccuracies underscored the pilot proficiency shortfall the VVS had been forced to endure as a consequence of four years of deprived funding for training. The inaccuracy of these deliveries resulted in many Russian losses to friendly fire. Most aircrews who participated in the initial attacks had not flown more than 30 hours the preceding year. Few were night-current or maintained any weapons delivery proficiency. As a result, General Deinekin was forced to assemble a "tiger team" from among his most experienced weapons instructors and test pilots. Only then did ground attack operations begin to show more effective results.

Russia's war in Chechnya—still resurgent at times—was emblematic of the security challenges the VVS is most likely to face in the decade ahead. The war was regional yet remote from the center of Russia. It featured a technologically unsophisticated yet almost fanatically determined ethnic opponent. It presented no air-to-air threat and offered a permissive environment for attacking aircraft other than at low altitude, where widely dispersed antiaircraft guns and shoulder-fired SAMs presented a constant threat. Finally, it entailed little by way of an opposing air force. Despite the occasional use of precision-guided weapons against key targets, quantity prevailed over quality in VVS operations.

There are few profound lessons from the war to be assimilated by the VVS, since so many of the problems dramatized by its spotty performance reflect nothing more than the funding shortage that has afflicted it since the USSR's collapse. The chief revelations were simply worst fears confirmed about the VVS's eroded capabilities as a result of financial starvation. The extent to which the VVS was strapped in fulfilling its tasking even in the relatively low-intensity war in Chechnya indicates that as long as it remains financially deprived, it will constitute, at best, only a regional air arm with little sustainability or capacity for high-technology combat.

THE OUTLOOK FOR A NEW RUSSIAN FIGHTER

One of the most intriguing questions about the VVS concerns the near-term likelihood of its developing and deploying a new air superiority fighter to replace the MiG-29 and Su-27. There is no doubt that Russia has both the means and incentive to develop such a fighter. Indeed, the Mikoyan Design Bureau could test-fly a prototype at any time if the repeated promises of its leaders can be taken at face value. However, Russia lives in a world of severe fiscal constraints that militate strongly against the full-scale deployment of a next-generation fighter in the class of the USAF's F-22 any time soon.

To begin with, Russia faces no security challenge that would even remotely warrant the expenditure of large amounts of scarce funds on the VVS's stated need for a new air superiority fighter. The main thrust of Russia's defense policy is toward power projection, not high-technology air warfare. As for the outlook for new acquisitions, First Deputy Defense Minister Andrei Kokoshin has said that available resources will permit little more than providing the armed forces with a minimal amount of new equipment for at least the near term.

Second, acquiring a new air superiority fighter is not the VVS's most pressing concern. General Deinekin has declared that the main goal of VVS modernization to the year 2000 is the creation of a mobility capability to support Russia's peacekeeping needs around its conflicted periphery. He has also said that the acquisition of new transport aircraft is the VVS's most urgent need.

Third, the VVS has other programs in train in the air-to-air mission area that promise attractive returns at a fraction of the cost of a new fighter. These include the Su-35 (a substantial improvement on the Su-27) and an active radar missile comparable to the American AIM-120 AMRAAM.

Fourth, whatever the VVS's stated requirements may be, it does not command the inside track when it comes to core decisions on resource apportionment. The VVS will never get a new fighter without the defense ministry's support. And by all indications, the Ministry of Defense has more urgent priorities. First Deputy Defense Minister Kokoshin has declared that the ministry's intended strategy is to concentrate on tried and proven equipment and to forgo programs that would not attain their projected specifications for years to come. This suggests that although the defense ministry will try its best to keep a modicum of sustainer funds channeled into Mikoyan's new fighter program, any serious development of the airplane will be subordinated, at least in the near term, to concentrating on improvement of the Su-27 and other fighters already in service.

Finally, even if the VVS and defense ministry could plead an airtight case for a new fighter, the question would remain: How will they pay for it? By late 1993, promised funding allotments had fallen so far behind that the defense ministry was a full trillion rubles in arrears to the defense industry for goods and services already delivered. That number rose to 2.2 trillion in 1994. The Ministry of Finance has consistently failed to authorize enough funds even to cover the VVS's fuel needs. As a result, continuation training in line units has dropped to crisis levels.

The VVS in 1993 received only 15 percent of the procurement funds it had been counting on, with no improvement in 1994. Even the manufacture of helmets and flight suits was cut off. General Deinekin admitted in 1994 that the VVS was doing its best to sustain Mikoyan's new fighter program but added that the lack of adequate financing was a major hindrance. For that reason, he concluded, it would be "not soon" when the most demanding new VVS projects come to fruition.

TRENDS AND PROSPECTS

In light of these problems, one might fairly ask whether the sun is rising or setting on General Deinekin's VVS. To this, General Deinekin would almost surely reply that his air force has the needed talent, an appreciation of its past failings under Soviet communism, a vision of what needs to be done to correct them, and an abiding conviction that, with patience, military aviation in Russia will eventually recover to full health. The problem is that the factors that will largely determine the course and outcome of the VVS's struggle for resurrection lie almost completely beyond General Deinekin's control. At bottom, the fate of the VVS, like that of the Russian military as a whole, is inseparably tied to the fate of post-Soviet Russia.

With respect to air doctrine and concepts of force employment, the VVS confronts a need to develop new approaches consistent with the emerging mission requirements of post-Soviet Russia. However, Russia has yet to develop a coherent foreign policy, or even an agreed-upon set of national interests upon which such a policy might be based. In the absence of a clear-cut threat or readily definable operational challenge, any attempt to produce a more detailed repertoire for Russian air power would amount to putting the cart before the horse.

In all, lean years lie ahead for military aviation in Russia, and much will turn on the still-uncertain prospect for economic stabilization and recovery throughout the country as a whole. Nevertheless, the VVS has entered a new phase in its evolution. The best of its new leaders—notably General Deinekin—have frankly admitted their many problems and have indicated what they believe needs to be done to begin fixing them. This has removed a major obstacle from the road to recovery. It has also set the stage for a time of creative ferment, once Russia emerges from its current crisis with a measure of fiscal solvency.

ACKNOWLEDGMENTS

Earlier versions of this report have been briefed widely to USAF and other audiences over the past three years. Most notably, these have included the Under Secretary of the Air Force; the Vice Chief and Assistant Vice Chief of Staff; the Deputy Chief of Staff for Plans and Operations; the Deputy Chief of Staff for Logistics; the Deputy Inspector General; the Director of Programs and Evaluation; and the Director of Plans at Headquarters USAF. Presentations were also made to the Vice Commander and Deputy Chiefs of Staff for Plans, Operations, Logistics, and Requirements at Headquarters Air Combat Command; the Deputy Chiefs of Staff for Operations and Plans at Headquarters Pacific Air Forces; the faculty of the Department of Professional Military Instruction, U.S. Air Force Academy; the 1993 Canadian, United Kingdom, and United States (CANUKUS) Air Power Conference; the 1994 Nellis Air Tactics Conference; the commanders of the USAF Weapons School, 4440th Tactical Training Group, and 422nd Test and Evaluation Squadron, Nellis AFB; the chiefs of staff of the Finnish Air Force and the Royal Australian Air Force; the commandant of the Royal College of Defense Studies; the director of the Air National Guard; the commander of the Air Force Studies and Analysis Agency; and the U.S. Air Attaché to Russia. The thoughtful comments and reactions provided by many of these audiences have helped sharpen the focus of the analysis presented in the following pages.

Written comments on earlier portions of this report were offered by Air Marshal John Allison, Chief of Staff, RAF Strike Command; Major Generals Lee Downer and Jeffrey Cliver, successive Deputy Chiefs of Staff for Operations, Headquarters United States Air Forces in

Europe; Lieutenant General Stephen Croker, Vice Commander, Air Combat Command; Mr. Steve Dunn, Headquarters Air Combat Command; Professor John Erickson, University of Edinburgh; General Richard Hawley, Commander, Air Combat Command; Lieutenant General Bradley Hosmer, USAF (Ret.), former Superintendent, U.S. Air Force Academy; Colonel Clifford Krieger, USAF, Department of Strategy, National War College; Air Vice Marshal Tony Mason, RAF (Ret.); Major General Donald Shepperd, Director, Air National Guard; Air Marshal Sir John Walker, RAF (Ret.); Brigadier General John Reppert, USA, Defense Attaché, and Major Dave Johnson, USAF, Assistant Air Attaché, U.S. Embassy Moscow; Major Randy Mayer, USAF (Ret.); my RAND colleagues Abraham Becker, Carl Builder, William O'Malley, and Milton Weiner; and a former Soviet fighter pilot now residing in the United States. I am indebted as well to my RAND colleague Ashley Tellis and to Rose Gottemoeller, Deputy Director of the International Institute for Strategic Studies, for their technical reviews of the final draft of this report, and to my editor, Malcolm Palmatier, for his usual keen eye. I owe a special word of acknowledgment to the Rockefeller Foundation for having granted me a five-week residency at its Study and Conference Center in Bellagio on Lake Como, Italy, where the initial outlining of this study was begun in September–October 1991.

Finally, I want to express my appreciation to General Petr Deinekin, commander in chief of the Russian Air Force, and to Major General Vasily Aleksandrov, head of the Russian Air Force's Central Research Institute, for their interest in this work and the opportunities they gave me to see them on numerous occasions during a succession of trips to Moscow since December 1991. I am likewise indebted to four top-ranked Russian test pilots with whom I have been privileged to fly in Russian airspace—Valery Menitskii, Vladimir Gorbunov, and Pavel Vlasov of the Mikoyan Design Bureau and Anatoly Kvochur of the Gromov Flight Research Institute—as well as to Rostislav Belyakov, head of the Mikoyan Design Bureau, Anatoly Belosvet, deputy general designer at Mikoyan, Grigory Sedov, Mikoyan's former chief test pilot, and Mikoyan pilots Roman Taskayev, Boris Orlov, and Marat Alykov for helping me to gain a more than bookish feel for the human side of military aviation in Russia. I have not, strictly speaking, "interviewed" any of these Russian air power principals in connection with my research, and none bears any responsibility for the findings of the present report. Nevertheless, my opportunity to be exposed to them worked greatly to eradicate any lingering doubt in my mind that Russian airmen stand proudly among the world's best as committed professionals.

I wish to dedicate this work to the memory of Colonel R. Moody Suter, USAF (Ret.), a fighter pilot's fighter pilot and idea man of rare talent and verve. Moody was the first to lend direction and focus to my interest in Soviet fighter aviation in 1976, and he remained my most abiding source of inspiration thereafter. A pioneer of realistic awareness of the Soviet fighter pilot and his operational style, he left a lasting mark on the way Western airmen do threat assessment today, and for the better. His spirited presence is greatly missed.

ACRONYMS

ADF	Automatic Direction Finding
ADI	Attitude Director Indicator
AFB	Air Force Base
AI	Air Intercept
AMRAAM	Advanced Medium-Range Air-to-Air Missile
AOA	Angle of Attack
ASF	Air Superiority Fighter
ASW	Anti-Submarine Warfare
ATC	Air Traffic Control
ATF	Advanced Tactical Fighter
ATGM	Antitank Guided Missile
CAF	Counterair Fighter
CBU	Cluster Bomb Unit
CFE	Conventional Forces in Europe
CIS	Commonwealth of Independent States
C3	Command, Control, and Communications
C3I	C3 and Intelligence
DACT	Dissimilar Air Combat Training
DNIF	Duty Not Including Flying
DOSAAF	Soviet Paramilitary Organization
ECM	Electronic Countermeasures
ERFA	Euro-Russian Fighter Aircraft
EW	Electronic Warfare
FAC	Forward Air Controller
FAIP	First-Assignment Instructor Pilot
FCF	Functional Check Flight
FCI	Fighter Combat Instructor
FOD	Foreign Object Damage

FOL	Forward Operating Location
FSK	Federal Counterintelligence Service
GAO	Government Accounting Office
GCC	Graduated Combat Capability
GCI	Ground-Controlled Intercept
GRP	Flight Supervision Group
GSFG	Group of Soviet Forces in Germany
IA/PVO	Air Defense Fighter Aviation
ICBM	Intercontinental Ballistic Missile
IFR	Instrument Flight Rules
ILS	Instrument Landing System
IP	Instructor Pilot
IOC	Initial Operational Capability
KBP	Combat Training Course
KCAS	Knots Calibrated Airspeed
KFA	Frontal Aviation Command
KGB	Soviet Committee for State Security
KP	Ground Command Post
LRA	Long-Range Aviation
MCM	Multi-Command Manual
MFI	Multirole Fighter
MVD	Ministry of Internal Affairs
NAS	Naval Air Station
NATO	North Atlantic Treaty Organization
NCO	Noncommissioned Officer
NII	Central Research Institute
NOE	Nap-of-the-Earth
NOTAM	Notice to Airmen
O&M	Operations and Maintenance
ORI	Operational Readiness Inspection
POL	Petroleum, Oil, and Lubricants
PGM	Precision-Guided Munition
QFI	Qualified Flight Instructor
RAF	Royal Air Force
R&D	Research and Development
RDT&E	Research, Development, Test, and Evaluation
ROE	Rules of Engagement
RP	Flight Director
RPG	Rocket-Propelled Grenade
RTU	Replacement Training Unit

SAM	Surface-to-Air Missile
SBP	Flight Safety Service
SSBN	Nuclear Ballistic Missile Submarine
TACEVAL	Tactical Evaluation
TCO	Tactical Control Officer
TECh	Intermediate Maintenance Section
UPT	Undergraduate Pilot Training
USAF	United States Air Force
USAFE	United States Air Forces in Europe
USSR	Union of Soviet Socialist Republics
VPVO	Air Defense Forces
VTA	Military Transport Aviation
VVAUL	Higher Military Aviation School for Pilots
VVS	Russian Air Force
WGF	Western Group of Forces

Chapter One

INTRODUCTION

In his masterful account of the Red Air Force's epic rise from near-disaster following the German onslaught in the summer of 1941 to its triumphant recovery four years later as the world's largest tactical air arm, Von Hardesty ably describes how this trial by fire had the ironic effect of providing Soviet combat aviation with "an accelerated passage to modernization and power."[1] In the immediate wake of the Soviet Union's implosion in December 1991, almost five decades after the start of Operation Barbarossa, the Soviet Air Force (*Voenno-vozdushniye sily*, or VVS) experienced an all too similar, if less apocalyptic, trauma. Almost overnight, it plummeted from its lofty status as a giant of some 20,000 pilots and 13,000 aircraft to become a new, and greatly impaired, organization of 13,000 pilots and only 5000 aircraft, mostly of obsolescent design.

To make matters worse, the VVS, like the other four services of the former Soviet Union, found itself possessed of few resources with which to sustain a prompt recovery. To this day, it remains embarked on an uncertain quest for renewed vitality and a new role in the post-Soviet and post-cold war world. For its current leaders, as for their predecessors in 1941, the crucial question concerns whether the blend of crisis and opportunity that circumstances imposed on them in 1991 portends a fate of inexorable decline or, in Hardesty's formulation, a situation from which they might engineer yet again "an accelerated passage to modernization and power."

[1]Von Hardesty, *Red Phoenix: The Rise of Soviet Air Power, 1941–1945*, Washington, D.C., Smithsonian Institution Press, 1982, p. 7.

Although it is not always appreciated in the West, military aviation has enjoyed a rich tradition in Russia. It predates and has thus far survived the 74-year intercession of Soviet communism. To note some of the high points: The world's first loop maneuver was performed by Major Petr Nesterov in 1913. Russia developed and successfully flew the world's first four-engine strategic bomber, Igor Sikorskii's *Ilya Muromets*, over the Eastern Front in World War I. Valery Chkalov commanded a pioneering flight in 1937 from Moscow to Vancouver via the North Pole.[2] Soviet airmen fought valiantly in World War II and played a key role in the defeat of Nazi Germany.[3]

The USSR led the way in jet aviation as well, with the introduction of the MiG-15 fighter in 1948. This was the world's first high-performance combat aircraft by modern standards, and it proved itself a technical match for the American F-86 in the skies over Korea.[4] In 1961, a Soviet fighter pilot, Yury Gagarin, became the first man to orbit the earth. Throughout the cold war, the VVS was uniformly taken by Western defense experts to be a formidable fighting force. By any measure, it and the Soviet aircraft industry, from their austere beginnings in the early 1920s to the enthralling flight demonstrations of the MiG-29 during its Western debut at the 1988 Farnborough Air Show, earned the USSR—and now Russia—legitimate pride of place as an international aviation giant.

Yet despite this rich background, most Westerners were only able to follow developments in Soviet military aviation from a distance until recently because of the Soviet leadership's obsession with secrecy and societal closure. The VVS was a denied area, a central component of the Soviet threat, and thus an object, first and foremost, of Western intelligence concern. Other observers had to view it darkly

---

[2]That event and its background are chronicled in Georgy Baidukov, *Russian Lindbergh: The Life of Valery Chkalov*, Washington, D.C., Smithsonian Institution Press, 1991. See also *V. P. Chkalov*, Moscow, Izdatelstvo Planeta, 1984.

[3]For a particularly engrossing memoir, see Marshal of Aviation Ye. I. Savitskii, *Polveka s nebom* ("A Half-Century with the Sky"), Moscow, Voenizdat, 1988.

[4]After the war, the U.S. Air Force test-flew a MiG-15 provided by a North Korean defector. Although the aircraft showed a number of undesirable handling characteristics compared with the F-86, it was found to be equally good, or better, in the critical parameters of maximum speed, service ceiling, and turn performance. The 10-1 kill ratio achieved by the F-86 over the MiG-15 in Korea reflected solely the superior training and proficiency of American pilots over their North Korean and Chinese opponents.

in an effort to understand what was going on beneath the often tantalizing, but rarely satisfying, appearances provided by the Soviet press and by periodic Western threat portrayals.

Because of this indistinctness, two noted British aviation experts pointed out in 1986 that "any attempt to describe the way aircraft are incorporated into the Soviet Air Forces, how they train, how they contribute to Soviet operational doctrine, and above all, how militarily effective they are, must be circumscribed again and again by conditions that apply in few other areas of military study." These analysts offered four good reasons why any bold assertions about the VVS needed to be advanced with studied care: "First, Russia goes to great lengths to conceal evidence of a kind which in the West may be found in technical journals, obtained from conversations, and observed on airfields. Second, much of the evidence which does become available is fragmented, sometimes contradictory, and frequently open to varying interpretations. Third, interpretation of that evidence, like any other, is susceptible to the preconceptions of the analyst. Finally, even if the evidence was comprehensive and the analysis always well judged and objective, the factors making up the equation of Soviet military effectiveness are so variable that a wide range of solutions would still be possible."[5]

That was sound counsel at the time it was written. Today, however, with the cold war over and the Soviet Union a fading relic of history, such obstructions to analysis have diminished substantially. Since the collapse of the USSR and the opening up of post-Soviet Russia, tracking the development of Russian military aviation has become considerably easier than it was during the darkest days of the cold war. The Russian media, including the military and technical press, have for the first time become purveyors of real facts rather than merely veiled hints in need of reading between the lines. More important, Russia's military and industry leaders have become accessible to their foreign counterparts and have shown a new willingness to engage in serious professional dialogue with Western defense experts. Given the current uncertainty regarding Russia's future political direction, these changes for the better may not prove to be last-

[5]Air Vice Marshal R. A. Mason, RAF, and John W. R. Taylor, *Aircraft, Strategy and Operations of the Soviet Air Force*, London, Jane's Publishing Company, Ltd., 1986, pp. 9–10.

ing. Nevertheless, throughout the five years to date since the USSR's collapse, they have made possible an unprecedented look at the VVS as it really is.

There remain limits, of course, that render Russia less than a completely open book. For example, Russia's recently dismissed defense minister, General Pavel Grachev, noted a few years ago that although already fielded military equipment is generally no longer treated as classified, future models and developments remain "state secrets."[6] Nevertheless, Russian military aviation is no longer a denied area or part of a declared military threat. More and more, as contacts between Western and Russian aviation professionals continue to grow, it is becoming possible to study Russian air power much as one would study military aviation in any other country.

This more benign atmosphere has revealed a Russian Air Force in the throes of a painful but determined metamorphosis. It is unmistakably embarked on a course of post-Soviet reform. Yet it remains uncertain of its future as it strives to embrace the 21st century as a renewed institution. Even before the demise of Soviet communism, there were gathering signs that the VVS, like the Soviet military as a whole, was entering its most turbulent time since its wartime trials in the early 1940s. Foremost among its challenges was adjusting to the radically changed setting of the post–cold war world at a time of deepening domestic political and economic crises.

For one thing, the end of the cold war and the collapse of the Warsaw Pact had left the VVS with no obvious threat and no clear mission beyond homeland defense. The "Warsaw Pact Air Operation," for which Soviet pilots and commanders had purportedly planned and trained for years, had become moot almost overnight. At the same time, the fact that the nation's defense was now obliged to begin at Russia's western edge meant that the VVS had assumed new responsibilities for which it was ill configured or prepared.

Beyond that, the freedom of expression made possible by President Mikhail Gorbachev's policy of glasnost introduced in 1986 had prompted an unprecedented venting of complaints throughout the

---

[6]Interview with General Pavel Grachev by V. Starkov, "General Grachev on the Military and on the Soldier," *Argumenty i fakty*, No. 5, February 1993, pp. 1–2.

armed forces. As a result, the VVS found itself besieged by new pressures from below. Among the openly aired problems with which its leadership now had to contend were declining aircrew morale and retention, an erosion in the quality and number of applicants to pilot training schools, and an appalling quality of life for aircrews and their families. There was also a growing acknowledgment of severe deficiencies in tactical air training, as well as mounting discontentment over the mindless administrative detail and paper-chasing that dominated squadron and regiment life, the pervasive lack of honesty and integrity within flying units, the poor quality and reliability of much of the equipment provided by the aviation industry, and increasingly apparent inequalities between their own and Western equipment and training.

The resultant criticisms hit especially hard on such perennial vexations as overly intrusive higher-headquarters meddling in day-to-day flight operations, seemingly endless paperwork and bureaucratic proceduralism at the regiment level, and the continuing tension between the demands of flight safety and the often conflicting imperatives of operational realism in peacetime training. Russian pilots and commanders registered unhappiness over more fundamental concerns as well, such as misplaced service priorities, rampant careerism and compromises of integrity by commanders looking mainly to "get ahead" within the system, and a consequent loss of vision and sense of purpose by the institution as a whole.

True enough, many of these complaints aired by Russian pilots and commanders since the onset of glasnost have sounded remarkably similar to those voiced for decades by professional airmen the world over. Indeed, they tend to bear out the popular belief that some simply go with the trade. As retired RAF Air Vice Marshal Tony Mason has aptly noted, "anyone who has spent any time in a barracks or squadron crew room, or as a senior officer has stayed on late at a unit dinner night, will know that British servicemen can hold their own internationally when criticizing the system, their equipment, their personnel management, and especially their senior commanders." The same goes for American or any other allied airmen. Accordingly, there is a natural tendency for senior officers in the West to brush aside such complaints as routine grousing without much practical import.

In the case of the VVS, however, as Mason rightly observes, the persistence of grassroots criticism of established norms since the beginnings of glasnost has been "too extensive, too specific, too repetitive, and too similar from many rank levels" for any such cautionary note to carry much weight. Mason further points out that behind many of these outcries from the ranks have been "a sense of hurt pride, frustrated professionalism, and a genuine desire for reform, rather than rejection or destruction."[7]

So what, one might ask? Why should anyone care any more about an air force that not only has ceased being a threat to Western security, but indeed finds itself operating in virtually a survival mode today? For one thing, the frank admissions of VVS pilots and commanders up and down the ranks since the beginning of glasnost give us an unprecedented chance to update and, where necessary, to correct our past impressions of the VVS. Beyond that, better knowledge of where the air arm of the former Soviet Union stands today can shed useful light on the course it may take once the current post-Soviet reform effort assumes a more even keel. Whatever difficulties and transition pains the VVS may be experiencing today, there is little doubt that Russia will eventually emerge from the USSR's collapse as a strong nation. There is also little doubt that the VVS, for better or worse, will constitute an important part of Russia's military capability.

This latter point takes on added strength in light of the striking similarity between many of the shared problems currently faced by the VVS and its most advanced Western counterparts. Among the most acute concerns confronting the VVS today are harnessing new technology to mission requirements at an affordable cost, enlisting and retaining high-quality people, keeping the defense industry accountable, sustaining the morale and motivation of VVS personnel at a time of major budget and force reductions, and ensuring that the VVS develops the organizational adaptability it will need to survive as a healthy institution in the 21st century. At bottom, these problems are not all that different from those facing most Western air forces today, the USAF included. However, they are compounded many

---

[7]Air Vice Marshal R. A. Mason, RAF (Ret.), *Air Power: A Centennial Appraisal*, London, Brassey's, 1994, p. 211.

times over in the case of the VVS by Russia's continuing political disarray and economic weakness. It should be instructive for Western planners to observe how their Russian counterparts are grappling with similar issues—and in a far more demanding situation than anything we have had to confront, at least yet.

This report begins with an overview of the predicament inherited by today's VVS leaders as they strive to extricate themselves from the depredations of 74 years of Soviet misrule and gain a firmer footing out of the turmoil of post-Soviet economic disorder. In particular, Chapter Two addresses such problems as the impact of budget cuts and other negative effects of the USSR's breakup, reduced fuel allocations to operational units and the resultant drop in annual flying hours for aircrews, the declining quality of life and need for better housing and amenities for VVS officers and their families, and other concerns confronting the VVS's leaders in their continuing effort to reestablish an acceptable level of institutional vitality.

Chapter Three examines declared VVS force modernization plans and the premises that underlie them. In so doing, it relies heavily on a detailed force assessment and projection recently prepared for open publication by the VVS's Central Research Institute. Among other things, this chapter describes the VVS's organization and inventory of aircraft and discusses VVS plans to reduce the diversity of its equipment and increase its emphasis on multirole systems.

Drawing heavily on the profusion of first-hand material that has appeared in the Russian military press on VVS operations and training since the beginnings of glasnost, Chapters Four, Five, and Six provide a detailed look at undergraduate pilot training and operational conversion, continuation training in line fighter units, and day-to-day organizational life at the regiment level. This spontaneous outpouring of information, unfortunately now noticeably reduced owing to a recent retightening of security clamps in Russia, has broken down much of the mystery that once shrouded all but the broadest outlines of Soviet fighter employment practice. In the process, it has highlighted a number of issues concerning fighter aircrew training, tactics development, and flight operations at the unit level that had long simmered but remained largely suppressed by the Communist Party's intolerance of dissent. As a result, questions that were hotly debated among Western analysts without resolution through the

1980s have finally been put to rest, confirming earlier suspicions that Soviet fighter aviation from at least the early 1970s onward was severely shackled by top-down restrictions on pilot initiative, independence, and adaptability. Consequently, Chapter Seven speculates with the benefit of hindsight about how the VVS might have performed in an aerial showdown against NATO, in the light of what we have since learned about its real strengths and weaknesses.

Chapter Eight assesses the VVS's role in the still-smoldering Russian incursion into secessionist Chechnya between late November 1994 and the end of June 1995. It first characterizes the nature and effectiveness of Russian air involvement in President Boris Yeltsin's badly botched effort to suppress the Chechen rebellion by force. It then examines the many problems highlighted by Russia's first combat test since the USSR's collapse, considers how the experiences and costs of the war are likely to affect the VVS's near-term recovery efforts, and reflects on what the performance of Russian air power in Chechnya has revealed about the health and capability of the Russian defense establishment as a whole.

Chapter Nine reviews the considerable evidence bearing on the near-term prospects for the development and production of a fifth-generation Russian air superiority fighter to replace the MiG-29 and Su-27. Despite a declared VVS requirement for such a fighter to match the USAF's planned F-22, and notwithstanding persistent assurances that the Mikoyan Design Bureau has a prototype in hand that it could fly at any time, the analysis argues that because of the continuing economic crisis that threatens the very livelihood of the VVS, the chances of any such fighter seeing deployment in enough numbers to make a difference before well into the next century are exceedingly remote.

Chapter Ten, finally, concludes with an overview of the VVS's predicament and reflects on possible future directions that Russian air power may take once Russia's national interests and foreign policy become more clearly defined. Its main message is that however honestly the VVS's leaders may have sized up their situation and however well prepared they may be to come to grips with it intellectually, any resurrection of military aviation in post-Soviet Russia will depend on factors that lie largely beyond their control. At best, like all the Russian armed services today, the VVS faces lean years and continued hard times.

Chapter Two

THE SOVIET LEGACY

The abortive coup attempt of August 1991 that started the clock ticking toward the fall of Soviet communism four months later affected the VVS much as it did the other Soviet services—and Soviet society across the board. With the old order roundly discredited and stripped of any lingering claim to legitimacy, the path was cleared for new looks at all aspects of the VVS's repertoire that had been driven by the idiosyncrasies of the Soviet state. At the same time, most remaining strictures against freedom of expression within the military were lifted. As the VVS's monthly magazine later commented, "glasnost continues to uncover an interminable stream of problems that used to be kept silent in the life of our armed forces."[1]

The first consequence of note for the VVS was a change in leadership. In July of the previous year, Colonel General Yevgeny Shaposhnikov had been appointed commander in chief to replace retiring Marshal of Aviation Aleksandr Yefimov, a veteran of World War II who had commanded the VVS since 1984.

That appointment marked a generational shift in the VVS's leadership. Considering that Shaposhnikov was picked over a large number of more senior officers, his appointment seemed to reflect a higher-level determination by the Gorbachev government to infuse the VVS with new blood. In contrast to the previous succession of commanders of Marshal Yefimov's vintage, Shaposhnikov typified a new breed of more technically minded officers who lacked the ideo-

[1]Colonel (Res.) V. Dudin, "Through a Mass of Stereotypes," *Aviatsiia i kosmonavtika*, No. 1, January 1992, pp. 4–6.

logical and historical baggage of his predecessors. Aged 48 at the time of his selection, he was (and remains) the youngest man to have commanded the VVS at any time in its postwar history.

From his first days as commander in chief, Shaposhnikov showed ample signs of being a reform-minded leader with little patience for the hidebound ways of the Soviet military bureaucracy. A fighter pilot by upbringing, he had commanded a MiG-23 regiment in the Group of Soviet Forces in Germany (GSFG) in 1975 as a lieutenant colonel at the young age of 34.[2] He subsequently climbed the VVS career ladder rapidly, advancing to the position of air commander for the GSFG in 1987 before being tapped to become first deputy VVS commander in chief the following year. Even Alexander Zuyev, the former VVS captain who defected to the United States via Turkey by flying a stolen MiG-29 from his base in Soviet Georgia to Trabzon on the Black Sea in May 1989, has attested that Shaposhnikov was well regarded among squadron pilots. Commenting on Shaposhnikov's role in putting down the 1991 coup attempt and his subsequent elevation to the position of defense minister, Zuyev characterized the VVS chief as "a real reformer, a patriotic professional officer who knew where his true loyalties lay."[3]

Among all the Soviet military chiefs, Shaposhnikov drew the line most forcefully against those on the High Command who had supported or otherwise sympathized with the plotters.[4] For refusing to abide the coup attempt, Shaposhnikov was selected to replace the disgraced General Dmitri Yazov as defense minister once the back of the putsch was broken.[5] Shortly thereafter, he was elevated to the rank of marshal of aviation. To take over his vacated post as VVS

---

[2]"Senior Air Force Appointments Confirmed," *Jane's Defense Weekly*, April 9, 1988, p. 705.

[3]This notwithstanding the fact that as first deputy VVS commander, Shaposhnikov reportedly headed the inquiry into Zuyev's theft of the MiG-29. See Alexander Zuyev, with Malcolm McConnell, *Fulcrum*, New York, Warner Books, 1992, p. 354.

[4]Later, Shaposhnikov claimed his readiness to attack the Kremlin had the putschists sought to take the White House by force. See Michael Evans, "Marshal Was Ready to Bomb Kremlin," *London Times*, September 13, 1991.

[5]Shaposhnikov presents a remarkably honest reflection on his experiences during the coup and the subsequent collapse of Soviet communism in his *Vybor: zapiski glavnokomanduiushchego* ("A Choice: Notes of a Commander in Chief"), Moscow, Nezavisimoye Izdatelstvo PIK, 1993.

commander in chief, he selected his first deputy, Colonel General Petr Stepanovich Deinekin.[6] In the immediate aftermath of the coup, both he and Deinekin resigned their memberships in the Communist Party of the Soviet Union.[7]

Unlike many previous VVS commanders, whose backgrounds had been with fighters, Petr Deinekin rose through the ranks of the Soviet bomber community. The son of a fighter pilot who died in 1943 while flying a LaGG-3 during the Great Patriotic War, he attended the Balashov Higher Military Aviation School for Pilots and eventually earned his wings as a bomber pilot, later serving on squadron duty and, in time, commanding a Tu-22M Backfire regiment. After that, he commanded an air division, an air army, and ultimately Long-Range Aviation (LRA) before being tapped by Shaposhnikov to become first deputy VVS commander in chief in 1990. During his career progression, he attended the Gagarin Air Academy and later graduated with honors from the General Staff Academy. He has more than 5000 hours of flying time, including an initial qualification checkout in the Tu-160 Blackjack. In June 1996, along with the other four Russian service chiefs, he was promoted to four-star rank by President Boris Yeltsin in a transparent bid for the support of the High Command just four days before the election.

What kind of new Russian Air Force, under General Deinekin's command, is now emerging from the wreckage of the failed Soviet system? Later chapters will explore VVS reorganization, force development, roles and missions, operations and training, and the intercession of a revealing combat test in Chechnya. This chapter is rather an initial damage assessment aimed at addressing those concerns that weigh most heavily on General Deinekin's mind as he seeks, above all, to ensure his air force's survival as an institution. Because of its high-technology orientation and the special demands on resources that this focus naturally entails, the VVS is possibly hurting more than other services from Russia's economic crisis. It bears noting, however, that much of what will be sketched out below regarding the VVS's post-Soviet tribulations can be said of the Russian

---

[6]See FBIS Eurasia report, September 3, 1991, p. 54.

[7]Interview with General Deinekin by N. Belan, "Gaining Altitude," *Sovetskaia Rossiia*, September 6, 1991.

armed forces as a whole. In this respect, General Deinekin is scarcely alone among his fellow service chiefs in the many difficulties he faces.[8]

INITIAL RESPONSES TO THE FAILED COUP

Not long after his appointment as defense minister, Shaposhnikov remarked that the August events had occasioned a "moment of truth for reviving the prestige of military service." He declared that the time had come to cast away, once and for all, the pervasive suspicion and distrust that had undermined the pursuit of combat readiness throughout the Soviet era. He added that the abortive coup had prompted long overdue decisions regarding the conduct of military life, and he promised sweeping changes based on the principles of professionalization, quality, democratization, and sufficiency.[9]

The first new measure announced by Shaposhnikov was the dismantling of the political control apparatus that had long plagued the full maturation of military professionalism in the Soviet armed forces.[10] This disestablishment of the military's Main Political Administration had immediate consequences for a large contingent of political officers in the VVS. Before the coup, the VVS had 8500 such officers, many of whom were pilots. Among these, 29 were generals.[11]

Shaposhnikov made departyization of the military a matter of professional pride. "It is a very important task," he declared, "and not just a sign of the times or of fashion. In a multiparty system, the military could become an object of contention. We need to be above that."[12] Once the political officers were defrocked, Shaposhnikov

---

[8]For more on this, see Benjamin S. Lambeth, "Russia's Wounded Military," *Foreign Affairs*, March–April 1995, pp. 86–98.

[9]"Revive the Prestige of the Army: Minister of Defense Marshal of Aviation Ye. I. Shaposhnikov Replies to Questions from a *Krasnaia zvezda* Correspondent," *Krasnaia zvezda*, August 31, 1991.

[10]"USSR Defense Minister's Appeal," *Krasnaia zvezda*, September 3, 1991.

[11]Interview by N. Belan, *Sovetskaia Rossiia*, September 6, 1991.

[12]Interview with Shaposhnikov by V. Izgarshev and A. Chernak, "I Follow My Conscience and the Law," *Pravda*, September 25, 1991.

said his next priority would be to professionalize the armed forces. He cautioned that such reform—"a major controversy until recently"—would not be attended to overnight, before the proper social impact studies and cost analyses were done. He did promise, however, that the problem would finally "get the attention it deserves."

Other reform goals declared by Shaposhnikov included redoubling efforts to address the gaping welfare needs of the armed forces, retiring a surfeit of unneeded generals, and easing restrictions on military secrecy. On the first count, Shaposhnikov warned that the most imposing problems, notably those pertaining to training and social needs, could not be righted in an instant. A more pressing concern, he said, was to "revive the concept of the honor of an officer and give it living content." Shaposhnikov reiterated his earlier charge that the rejected Soviet system had been responsible for maintaining a pernicious atmosphere of "strife, distrust, and confrontational relations" throughout the armed forces that was "ruinous to the interests of combat readiness."[13]

As for accumulated deadwood, Shaposhnikov spoke of "hundreds" of generals who had remained on the payroll too long. He suggested that perhaps only 15 to 20 percent of these would ultimately be retained.[14] He further called for a reduction of obligatory service from two years to 18 months and for the gradual displacement of mandatory conscription by a volunteer system.[15] Finally, with respect to dismantling the despised machinery of Soviet censorship, he

---

[13]Interview with Marshal Shaposhnikov, "Revive the Prestige of the Military," *Krasnaia zvezda*, August 31, 1991.

[14]Interview by B. Moseichuk, "Ye. Shaposhnikov: 'We Are Shedding Excess Weight,'" *Argumenty i fakty*, No. 38, September 1991, p. 2.

[15]As currently planned, conscription in Russia will not be eliminated entirely, both because an all-volunteer military would be prohibitively costly and because the General Staff wishes to maintain a mobilization capability that only a draft can guarantee. In an eleventh-hour campaign ploy to tap the youth vote, President Yeltsin recently promised that, if reelected, he would end conscription by the year 2000. Any such move, however, would encounter strong resistance, not only from the military leadership but also from a sizable cross-section of legislators in the State Duma. It has already been challenged by Yeltsin's new security adviser, Aleksandr Lebed.

declared a presumption of openness in all activities that did not entail legitimate military or state secrets.[16]

For his part, the new VVS commander in chief, General Deinekin, left no room for doubt about where he felt the blame belonged for the many years of stagnation in VVS practice. He said that "the processes of departyization and depoliticization . . . that have been initiated actively reflect the long-standing attitude of most military flyers. The party political structures that existed interfered constantly and quite persistently in the conduct of virtually all aspects of our combat training, tying the hands of commanders and specialists."[17] Asked later what he felt the effects would be on the VVS as a result of the dismantling of these structures, General Deinekin replied: "Regardless of the final shape the reform will take, the air force will benefit from it. This country's air force suffered the burden of communism for 74 years. Now that burden has finally ceased to exist."[18]

IMMEDIATE CONSEQUENCES OF THE BREAKUP OF THE USSR

Marshal Shaposhnikov identified a number of destabilizing by-products of the disintegration of the Soviet Union: (1) an epidemic of nationalism in some former republics, (2) territorial claims due to past arbitrary determination of borders between republics, (3) a desire among many former republics to eliminate the "imperial center," (4) friction between former republics aggravated by the incomplete di-

---

[16]For background, see Stephen Foye, "Evgenii Shaposhnikov: A New Defense Minister for a New Era," *Report on the USSR*, Radio Liberty Research, September 31, 1991, pp. 8–11.

[17]Interview by N. Belan, "Gaining Altitude," *Sovetskaia Rossiia*, September 6, 1991. General Deinekin recalled how in earlier days the VVS might wish to advance a promising young officer, only to hit a brick wall in the party's Central Committee, where "some young man in a gray suit and blue shirt would pick up a special telephone and say that this candidacy did not suit them."

[18]Quoted in Alexander Velovich, "Soviet Forces Face Restructure," *Flight International*, September 25–October 1, 1991. Later, Deinekin added with pride that "the spirit of democracy is inherent in aviators." Interview by Yelena Agapova, "The Skies Are the Same—the Concerns Different," *Krasnaia zvezda*, October 26, 1991. He also affirmed during a press interview in Germany that the leadership in Moscow had broken with communism once and for all. Rudiger Moniac, "All Commitments Regarding Germany Will Be Precisely Honored," *Die Welt*, December 12, 1991.

vision of the spoils of the former USSR, (5) lack of economic stability, and (6) the presence of Russian forces outside of Russia.[19]

In the face of these disruptions and dislocations, the newly installed Russian government of President Yeltsin took determined steps to supplant the USSR with a Commonwealth of Independent States (CIS) that might preserve at least a semblance of the former union's integrated defenses. A CIS Joint Armed Forces command was created, and Marshal Shaposhnikov was named its commander in chief. At best, the CIS faced an uphill climb from the very beginning because of the refusal of most former Soviet republics, notably Ukraine, to sacrifice their newly gained sovereignty to this new and suspect entity.

Problems with Air Defense

The collapse of the Soviet multinational state posed an immediate challenge to the air defense of the former USSR's territorial space. Almost alone among modern military powers, the Soviet Union—and, after its disintegration, Russia—vested this critical function in a separate service, the Air Defense Forces (*Voiska protivovozdushnoi oborony*, or VPVO). Because air defense is such a crucial part of Russia's military aviation complex and because of VPVO's close relationship with the fighter arm of the VVS, it will be given due attention here despite the study's main focus on the VVS.

Even before the disintegration of the union, the newly appointed VPVO commander in chief, Colonel General Viktor Prudnikov, conceded that the echeloned and multitiered Soviet air defense system was overdue for a top-to-bottom review now that the cold war was over and the NATO threat had ceased to exist.[20] He stressed, however, that the system must remain unified regardless of whatever changes might be put into effect. Anticipating the USSR's impending breakup, he rightly cautioned that no single state in a renewed con-

---

[19]Marshal Shaposhnikov, "National and Collective Security in the CIS," *Krasnaia zvezda*, September 30, 1992.

[20]General Prudnikov, former commander of the Moscow Air Defense District and a respected fighter pilot, was chosen to replace Army General Tretyak as commander in chief of VPVO after the latter was fired for having backed the coup attempt.

federation would possess the needed wherewithal to create such a system independently. "Any fragmentation of aerospace defense forces," he pointed out, "whether in terms of republics or in terms of branches of the armed services, will inevitably result in considerable expenditure toward developing parallel command structures. That will be costly, especially today, and ineffective."[21]

In a clear challenge to the idea of continuing VPVO as a separate service, especially in the wake of the end of the cold war, Army General Vladimir Lobov, who had been appointed chief of the General Staff after the failed August coup, raised an eyebrow at "whole clusters of duplicative parallel military structures," singling out in particular "air units in the VVS and VPVO that perform similar, if not identical, functions."[22] Later, General Prudnikov granted that a reassessment of VPVO's existing organization might be warranted, although he clung insistently to the rock-bottom need to maintain its separate and distinct command and control system.[23]

The chief of the VPVO headquarters staff, Colonel General Sinitsyn, amplified on this, citing various reorganization proposals that had come to light, including the distribution of VPVO assets among other branches of the military. He said that this option harked back to the attempted reforms of 1980, which had prompted a reduction in the combat capability of VPVO units and a dilution of accountability for executing the air defense mission. General Sinitsyn added that in 1986, "having realized that this imposed decision was in error, we were forced to return to a unified VPVO command, having wasted

---

[21]Interview with Colonel General V. Prudnikov, "A View of the Problem: What Should Aerospace Defense Be Like?" *Krasnaia zvezda*, November 30, 1991. In 1987, some prominent Soviet civilian analysts, notably Aleksei Arbatov, had proposed merging VPVO aviation with the VVS. See Arbatov's "How Much Defense Is Enough?" *Mezhdunarodnaia zhizn*, No. 3, 1989, pp. 41–43, and his later "Toward the Question of Sufficiency in Air Defense," *Voennaia mysl*, No. 12, 1989, pp. 41–45. That suggestion was roundly rejected by the High Command.

[22]"Lobov Assesses Military Reforms, Ramifications," *Krasnaia zvezda*, September 10, 1991.

[23]For background discussion, see John W. R. Lepingwell, "Gorbachev's Strategic Forces Initiative: Dissolving the Air Defense Forces," *Report on the USSR*, Radio Liberty Research, December 6, 1991, pp. 4–9.

considerable resources and lost some personnel during the interim."[24]

In light of this, Sinitsyn continued, it was curious that some people were yet again proposing to abolish the unified aerospace defense system: "Such a move would cause serious problems for the command of VPVO forces and personnel. The air defense system that has been built up, consisting of an array of interdependent, truly unique, and very costly systems, will fall apart." General Sinitsyn volunteered that VPVO's leadership had developed a plan that included integrating elements of VPVO into a new service to be called the Strategic Deterrent Force. This, he suggested, "would help preserve the unified air defense system that was formed over a period of decades and would prove its value in the new political conditions."

General Prudnikov was even more explicit regarding the need for a unified air defense net for the embryonic Commonwealth of Independent States. He pointed out that air defense of the CIS's western portion had "substantially worsened" as a result of the collapse of the Warsaw Pact. This meant the loss of a forward radar zone covering a depth of 800–1000 km, to say nothing of a changed disposition of alert VPVO interceptors and surface-to-air missiles. General Prudnikov warned that the problem would become even more critical after the withdrawal of VPVO units from the Baltic republics. He spoke with guarded hope about the prospects for coordinating CIS air defenses with the national systems of Eastern Europe on a bilateral basis, citing a protocol signed with Romania as a promising step in the right direction. However, he added, "we really need to rely on our own forces."[25]

The head of VPVO's Center for Operational and Tactical Research likewise remarked that the Soviet High Command had spent years building a layered air defense with the best equipment and technical experts, only to be confronted today with a major breach in its former front line, with Ukrainian leaders demanding that former Soviet VPVO assets on Ukrainian soil be handed over for their own

---

[24]Interview with Colonel General V. Sinitsyn, "How Many Duplicates for the Keys to Heaven Are There?" *Krasnaia zvezda*, December 24, 1991.

[25]Interview with Colonel General Viktor Prudnikov, "Rely On Our Own Forces," *Armiia*, No. 24, December 1991, pp. 7–11.

sovereign use. Unless such "anarchy" were soon brought under the guiding control of common sense, he warned, "things may reach a point where it will not be just Rust who will fly in."[26]

What seemed at first to be a step toward retaining at least a modicum of integration occurred on May 26, 1992, when the first deputy Ukrainian defense minister signed an agreement in Moscow assuring Ukrainian participation in a joint system of antimissile defense and in coordinating air defense.[27] This was short-lived, however; the following October, the CIS armed forces leadership was said to be "making desperate efforts . . . if not to preserve the air defense system in its old form, at least to organize reliable cooperation among its republics in this sphere."[28] At a conference of CIS defense ministers in Bishkek, Shaposhnikov stressed the urgency of an agreement on unified air defense. But the idea never got past the talking stage.

In mid-1992, General Prudnikov reported that at least some of the unified air defense architecture of the former USSR had been preserved, although efforts by various CIS member states to nationalize their portions threatened to undermine the whole system. "Practically every day," he complained, "we experience bans on the movement of our units, or a disruption in the regular flow of trains to be offloaded of equipment, or problems with pay. . . . These are not isolated cases."[29] On the positive side, Prudnikov noted that although VPVO had lost some of its westernmost borders as a result of the union's collapse, radar coverage remained multiechelоned and effective.

The commander of the Moscow Air Defense District, Colonel General Kornukov, also stressed that the idea of an echeloned and perimeter defense must remain central to VPVO's concept of operations. He

---

[26]Interview with Colonel Viktor Demediuk, "It May Not Be Just Rust Who Will Fly In," *Pravda*, January 11, 1992. The reference is to Matthias Rust, the West German teenager who scandalized the Soviet defense establishment in 1988 when he penetrated the USSR's western frontier in a Cessna 172 and landed unmolested in Red Square.

[27]Comment by Marshal Shaposhnikov in an interview with Colonel V. Litovkin, "A New Chance," *Priorities and Prospects*, Moscow, No. 3, September 1992, pp. 15–19.

[28]Aleksandr Stukachin, "The CIS Air Defense System Is Disintegrating. Shaposhnikov Is Trying to Save the Air Shield," *Kommersant*, October 16, 1992.

[29]Moscow television First Program, July 20, 1992.

said he was not particularly worried over the loss of the all-union unified system, although he admitted that the loss of the forward area had had a deleterious impact on early detection and warning. In effect, this meant that the Moscow Air Defense District had now become a frontier command, obliging its personnel to abandon any complacent notions of still being safely in "the rear."[30]

By the end of the year, Prudnikov was no longer skirting the issue of the effect of the union's collapse on the air defense of Russia. Asked how things had changed during the preceding year, he answered: "For the worse. The process of sovereignization has deprived the integrated air defense system of many of the components that ensured its reliable functioning."[31] He also had little good to say about gathering calls to "reform" VPVO: "As I recall, and I have over 32 years of service in VPVO, there is always someone bent on transferring and splitting up the forces. The most recent attempts of this kind took place in the 1980s, when air defense units were transferred to the [military] districts. Analysis showed that aside from damaging national security and causing problems, this 'reform' produced nothing. On the contrary, the prestige of VPVO was undermined and, along with it, its combat potential."

General Prudnikov argued that such moves invariably produced additional layers of management, with the net result that each new tier "means lost time. If that happens, it means either a wrong decision is made, or the decision is made too late." Citing the September 1983 Soviet downing of Korean Airlines Flight 007 as a good example, he said that "we should have straightened things out the very moment the aircraft penetrated our air border, not at the last minute. Perhaps then the tragedy could have been averted."

---

[30]Interview with Colonel General A. Kornukov, "The Eyes and Ears of the 'Zone of Continuous Operations,'" *Moskovskaia pravda*, July 11, 1992. General Kornukov took special pains, however, to stress that he saw no designated threat. As for who the likely enemy might be, he stressed that labels like "aggressive U.S. imperialists," "German revanchists," and the "cynical British" all "come from the realm of political hypocrisy. This is unprofessional. . . . 'Violator' is an expansive enough term to define my attitude toward a stranger."

[31]Interview with Colonel General Prudnikov, "Will the Likes of Rust Get Though Again?" *Rossiiskiye vesti*, December 12, 1992.

General Prudnikov further disclosed that only some 70 percent of VPVO's original assets remained on Russian soil, and that it would take time to get Russian air defense integrity back to an acceptable level. Despite lip service from other republics for an integrated surveillance and monitoring system, these same republics declined to recognize the operational control of Marshal Shaposhnikov's CIS joint command. In some cases, notably in Transcaucasia, they also lacked the trained manpower needed to operate the radar stations.

Thus a strong initial Russian hope that a unified CIS air defense might be preserved eventually proved to be unfounded. On the eve of the breakup of the USSR, General Prudnikov warned that any fragmentation of air defenses among the CIS republics would impose considerable costs on Russia to create new command structures in their place.[32] In short order, that is exactly what happened. The result was an almost overnight disappearance of what Marshal Shaposhnikov called "a single military-strategic area" developed over a 70-year span of Soviet history.[33]

Today, Russia's air sovereignty is protected mainly by the surviving remnants of the former Soviet VPVO, with some of the resultant slack having been taken up by the fighter arm of the VVS.[34] The defense ministry's declared plans for the third stage of an ongoing military reorganization implied that VPVO might eventually be disestablished, with its core assets being absorbed into the VVS and a newly constituted Strategic Deterrent Force. For the moment, however, VPVO remains an active part of Russia's military aviation complex, and its operational training continues despite resource restrictions, undermanning, and "chronically belated funding."[35] Although its

---

[32]Interview with Colonel General Prudnikov, "A View of the Problem: What Should Aerospace Defense Be Like?" *Krasnaia zvezda*, November 30, 1991.

[33]V. Litovkin, "The Army Is Ready to Obey the Presidents: The Presidents Are Trying to Agree Among Themselves," *Izvestiia*, December 31, 1991.

[34]See "The New Russian Air Force: Interview With the Commander in Chief of the Russian Air Force," *NATO's Sixteen Nations*, No. 3, March 1993, p. 39.

[35]Colonel General Kornukov reported that half his interceptor regiments and all his surface-to-air missile regiments were tested on the range in 1993, with an overall "firing efficiency" of 95–96 percent. That claim strains credulity, considering the multiple constraints that by then had come to afflict VPVO and VVS units of all types. Nikolai Poroskov, "The Air Defenses of Russia's Central Region Have Been Checked. The Results Are Reassuring," *Krasnaia zvezda*, November 30, 1993.

future composition and ultimate prospects for survival as a separate service remain to be determined, the suggestion in 1992 that it "will soon be extinct" was premature.[36] To cite one recent indication of the reluctance of many in the Russian military to let go of old ways, retired army Lieutenant General Aleksandr Lebed, a moderate nationalist contender for president during the 1996 campaign and now President Yeltsin's national security advisor, has advocated retention of the existing five-service arrangement, at least for the near term.[37]

Impact on the Defense Industry

Serious problems were portended as well by the dispersion of the Soviet aircraft industry as a result of the USSR's breakup. Early on, General Deinekin warned that the aviation industry had become "so interwoven with the VVS" that splitting it up among different republics "could endanger the very existence of the armed forces." The good news was that about 85 percent of the plants of the former Soviet military aircraft industry remained on Russian soil.[38] Nevertheless, in the interest of maintaining an integrated CIS air capability, General Deinekin said he was hoping to negotiate effective horizontal ties between various regions and republics, notably Ukraine, the second most important after Russia for aircraft production and the main supplier of transport aircraft and engines of all types.

The VVS also inherited serious supply problems as a result of the collapse of the USSR. Colonel General Anatoly Malyukov, the chief of the VVS headquarters staff, reported that even in late 1991 the situation had already become very serious. Among other things, he said that the VVS had had to make do without new batteries for a time because all aircraft batteries were manufactured in Ukraine.[39] The

---

[36]Major Brian Collins, "Russia Fragments Its Airpower," *Air Force Magazine*, February 1992, p. 62.

[37]For more on Lebed's possible influence in redirecting the various military reform proposals discussed in this report, see Benjamin S. Lambeth, *The Warrior Who Would Rule Russia*, Santa Monica, California, RAND, MR-805-AF, 1996, forthcoming.

[38]TASS report on a comment by General Deinekin, March 27, 1992.

[39]Quoted in Piotr Butowski, "Flying in the Face of Adversity," *Jane's Defense Weekly*, April 17, 1993, p. 19.

Lugansk factory ceased shipping batteries, and the Baltic states halted the production and shipment of aircraft radios.

In commenting on Ukraine's claim to all former Soviet assets deployed on its territory, Marshal Shaposhnikov pointed out that Ukraine lacked even a single combat aircraft production facility. In light of this, he predicted that "aircraft stationed in Ukraine will cease flying within six months at the outside. There will be no engines, spare parts, or tires. At the same time, there is no mechanism through which these commodities can be supplied from Russia, either by barter or for hard currency."[40] General Deinekin described the situation as the inevitable result of a conscious Soviet decision, harking back to Stalin's time, to set up aircraft design bureaus in some republics, manufacturing plants in others, and engine factories in still others—all toward the goal of ensuring the economic integration of the multinational Soviet state.[41]

As CIS joint forces commander, Marshal Shaposhnikov expressed Russia's readiness to help member states set up their own indigenous armed forces, including a system for acquiring weapons, equipment, spare parts, repair systems, and the required training.[42] He predicted in early 1992 that, without this, efforts by members to break up the integrated former Soviet VVS would inevitably lead to a shortage of fuel and spares for republican air forces, since aircraft were assembled and spares were manufactured largely in Russia and since there was no mechanism for selling spare parts to the republics if they should establish their own air forces.

On the domestic front, the VVS deputy commander for logistics, Lieutenant General Stanislav Ivanov, indicated growing concern over the potential loss of "responsibility, discipline, and order" in relations with the VVS's Russian suppliers.[43] In particular, General Ivanov complained that the flow of supplies was being disrupted by "new economic relations" and that deliveries were no longer being

---

[40]Quoted in N. Belan, "When We See Clearly It Will Horrify Us: Notes from Marshal of Aviation Ye. Shaposhnikov's News Conference," *Sovetskaia Rossiia*, January 7, 1992.

[41]Moscow television First Program, January 15, 1992.

[42]Moscow television First Program, January 17, 1992.

[43]Interview with Lieutenant General Stanislav Ivanov, "The Rear Is Becoming the Front as Well," *Aviatsiia i kosmonavtika*, No. 9, September 1991, pp. 2–3.

provided in an orderly manner through Gosplan, but rather on the basis of contracts negotiated with individual producer plants. "Not all suppliers," he said, "will meet us halfway. We frankly don't know which is better—for the state to maintain a monster like Gosplan, which prescribed everything for everybody, or for VVS rear services and enterprises to keep a special staff that travels around the country scrounging whatever they need." As for the impact of the "new business conditions" inspired by Gorbachev's economic reforms, General Ivanov lamented: "We are not feeling any advantages whatsoever in anything."

From the CIS to Russian Unilateralism

In May 1992, the Russian armed forces were born for the second time in modern history in the wake of vain efforts by the Yeltsin government to establish a viable CIS military organization. As early as two weeks after the breakup of the USSR, there was speculation in Moscow that, should negotiations toward a unified CIS prove unsuccessful, there would be every reason for Russia's leadership to declare Russia the USSR's successor in military matters.[44] Not long afterward, the February accords signed by each CIS member in Minsk gave each republic the right to create its own armed forces.[45]

Upon assuming his role as commander in chief of the CIS Joint Armed Forces, Shaposhnikov rued the fact that the profession of arms in the USSR had been forced to endure "the grief of the Afghan war, the pain of internal feuds, and an insulting lack of understanding by society, along with undeserved reproaches, instant poverty, and a lack of social prospects."[46] A priority goal of the new Russian leadership, he said, was to undo these corrosive influences as quickly as possible. Ultimately, that challenge fell to Yeltsin's new defense minister, Army General Pavel Grachev. However, Marshal

---

[44]Pavel Felgengauer, "The Army Is Under Fire From the Politicians," *Nezavisimaia gazeta*, January 11, 1992.

[45]See the interview with Shaposhnikov in *TASS-Skript*, No. 62, March 23, 1992.

[46]Moscow television newscast, January 17, 1992. He later remarked, on the eve of the failed attempt in Minsk to reach agreement on a unified CIS defense arrangement: "Ultimately I dream of a military that does not have to think about where its next meal is coming from." Moscow television newscast, February 13, 1992.

Shaposhnikov did much to pave the way in the immediate wake of the failed August coup. During his four months as the USSR's last defense minister, he eliminated Communist Party influence from the armed forces, abolished party control structures in the military, disestablished the so-called "paradise group" of inspectors in the defense ministry—a sinecure for semi-retired marshals—and discharged some 700 unneeded generals into the reserves.[47]

TROUBLED RELATIONS WITH UKRAINE

Russia's dealings with Ukraine soured almost from the first moments following the disintegration of the Soviet state, resulting in considerable part from an ownership dispute over the plenitude of front-line military hardware that Ukraine inherited from the former Soviet Union. Because of the USSR's western strategic orientation, much of its best and most modern combat equipment had been fielded on Ukrainian soil. As a result, Ukraine emerged from the union's collapse as, among other things, the possessor of over 1000 military aircraft, including between a quarter and a third of the former Soviet Air Force's MiG-29s and Su-27s; half of its forty Il-78 tankers; almost half of its Il-76 transports; and all but three of its serviceable Tu-160 Blackjack bombers. These assets instantly endowed Ukraine with an air force considerably larger than that of any West European country, including Britain, France, and Germany.

The Ukrainians have been touchy about their claimed proprietary rights to the former Soviet VVS equities left on their territory. In response to General Deinekin's warning that the Tu-160s based at Priluki would become inoperable if Ukraine did not promptly release them to Russia, a Ukrainian defense ministry press release countered that the aircraft "are in working condition and are completely ready," and that Ukrainian pilots are flying them and "are not losing their

---

[47]Interview with Marshal Shaposhnikov, "The Russian Army's History Did Not Start Today . . .," *Izvestiia*, May 8, 1992. In one sense, doing away with political officers amounted to throwing out the baby with the bathwater. Many were pilots, occasionally even respected ones. They fulfilled important roles that went beyond political indoctrination, such as the sort of counseling of junior officers that might be provided by a chaplain or social worker in Western military establishments. Accordingly, the plan was not to force these officers out entirely, but rather to redefine their functions and put them under the purview of a depoliticized education and training administration.

skills."[48] This implausible claim added that there was no disputing Ukraine's "privatization" of these aircraft, since "it is a matter for a sovereign state to decide how, when, and what it will do with its property."

The first outright sign of a gathering standoff between the two new countries occurred in early February 1992, when six Su-24 interdiction aircraft were secretly flown from their base in Ukraine to Russia by defecting Russian aircrews who refused to swear an oath of allegiance to Ukraine. CIS armed forces commander Shaposhnikov rejected a demand from Ukraine's president at the time, Leonid Kravchuk, that the aircraft be handed back to Ukraine and that the aircrews be returned to stand trial for desertion.

As a result of this episode, Ukraine unilaterally announced on February 17 that it was nationalizing the CIS air division at Uzin, consisting of a regiment of 22 Tu-95 Bear bombers, an Il-78 tanker regiment, and a support transport squadron. Two days earlier, with Shaposhnikov's blessing, the commander in chief of CIS Long-Range Aviation, Colonel General Igor Kalugin, had fired the division's commander, Major General Bashkirov, for having sworn an oath of allegiance to Ukraine. Bashkirov was promptly reinstated by Ukraine's then–defense minister (and General Deinekin's former VVS colleague), Colonel General Konstantin Morozov. This contretemps followed earlier bad feelings triggered by an effort by Moscow to order a number of Il-78 tanker crews based at Uzin to deploy with their aircraft on a CIS "training mission" to Russia. Suspecting a Russian ploy to gain physical possession of the aircraft and so to claim ownership of them, the crews refused and were backed by Bashkirov.

Earlier, the division in question had become riven with controversy over whether to yield to a demand by Kiev that its officers swear an oath of allegiance to Ukraine. The tanker regiment's aircrews took the oath hastily, at night, and under duress. Most of the bomber crews refused, with the predictable result that the division became split. Without tanker support, the operational reach of the bomber

[48]Holos Ukrainy report, March 27, 1993. This comment came in response to an interview with General Deinekin, "The Russian Air Force Will Have Only the Most Up-to-Date Aircraft," *Izvestiia*, March 24, 1993.

division became considerably reduced. As General Deinekin later commented wryly: "Pardon my unparliamentary language, but bombers without tankers are like eunuchs."[49]

Ukraine's insistence on Russian respect for Ukrainian sovereignty escalated sharply on March 21, 1992, when General Deinekin, in his then-role as CIS air forces commander, sought to fly to Ukraine to inspect CIS strategic flying units. The flare-up began with General Deinekin's sending a coded message to General Morozov stating his intended route plan and visit schedule for five stops in Ukraine beginning on March 24. Kiev, in turn, dispatched an icy reply forbidding General Deinekin to "fly over the airfields of a neighboring state." Its message added that since Moscow had not reached agreement with Ukraine on the status of CIS strategic nuclear forces on Ukrainian soil, "your [i.e., General Deinekin's] presence in Ukraine's armed forces is not expedient."[50] In yet a further escalation, CIS commander Shaposhnikov fired off an angry demarche accusing the Ukrainian defense ministry of violating CIS accords and upbraiding Kiev for obstructing Deinekin in the legitimate performance of his CIS duties. The denouement, on March 25, saw Kiev finally rescind its initial denial and grant approval for Deinekin to visit Ukraine as planned.

General Deinekin later received a group of Long-Range-Aviation (LRA) pilgrims from Ukraine who had declined to repudiate their allegiance to the former Soviet Union.[51] These Russians had been given an ultimatum either to sign an oath of loyalty to Ukraine or to vacate the republic's territory. General Deinekin assured them that Russia needed their experience and devotion to flying. He said that altogether, 690 Russian airmen had returned from Ukraine after having been forcibly removed from flight status and deprived of living accommodations.

---

[49]A. Krainiy, "A Chronicle of Mutinous Bombers: What Politicians Had Long Been Seeking Has Happened—the Military Has Split," *Komsomolskaia pravda*, March 27, 1992.

[50]Aleksandr Anin and Vladimir Svartsevich, "Conversion Commerce and Combat Training: The Division Commander Did Not Wish to Speak to Your Air Force Commander," *Nezavisimaia gazeta*, March 27, 1992.

[51]Lieutenant Colonel A. Aleksandrov, "Russia Will Not Abandon Its Sons in Their Hour of Need," *Krasnaia zvezda*, July 2, 1992.

If anything, Ukraine's air force was in even worse straits than Russia's after the disintegration of the union because of rampant supply problems. Only two flight turns were generated at the LRA base at Uzin for two full months in early 1992. Some pilots had gone three or four months without flying.[52] At the fighter base at Voznesensk, a MiG-29 regimental commander reported that the rate of training in his unit was "substantially less" than the two or three times a week he flew in the Soviet Air Force.[53]

Shortly after assuming command of the newly independent air arm, Lieutenant General Valery Vasiliev declared that problems with access to fuel, engines, spare parts, and repair were not tied to Ukraine's declaration of independence or to any troubled political dealings with Russia.[54] Rather, he insisted, they were the result of broken economic ties between now-separated production entities and an associated drop in production. Vasiliev claimed that the Ukrainian VVS's operational status was on a rough par with Russia's. He added that because of some regimental and squadron commanders' decisions to return to Russia, vacancies had opened up, especially for squadron commanders. He noted that the Ukrainian air force would continue to experience a drawdown in units. He also announced the establishment of four new operational headquarters, one on the western sector, in Lvov; another on the southwestern sector, in Odessa; and finally a headquarters each for transport aviation and for reserves and training.[55]

A subsequent report sounded considerably less upbeat. A Ukrainian VVS captain complained that deliveries of aviation fuel had recently been reduced to a minimum, with the result that "flight personnel are seizing any chance to fly to forestall a break in their training and

---

[52]Major S. Prokopenko, "Whose Side Are You On, Guys, Who Will Lead You Into Battle?" *Krasnaia zvezda*, May 20, 1992.

[53]Douglas Jehl, "Ukraine: A Nuclear Power, But Untested Loyalties," *New York Times*, December 2, 1993.

[54]Formerly commander of the Soviet Air Force's Vinnitsa Air Army, General Vasiliev voiced his intent to model Ukraine's air force along the lines of the Vinnitsa Air Army, with headquarters in Vinnitsa. In effect, this amounted to little more than a name change from the Vinnitsa Air Army to the Ukrainian VVS. See Nikolai Baras, "Air Forces Organized in Vinnitsa," *Pravda Ukrainy*, May 6, 1992.

[55]Interview with Lieutenant General Valery Vasiliev, "Aviators Have Chosen to Serve the People of Ukraine," *Narodnaia armiia* (Kiev), June 2, 1992.

to avoid losing their proficiency."[56] This captain added that cannibalization had become common owing to a "chronic" shortage of spare parts, even though the maintenance experts "know full well that they should not remove equipment from one aircraft to another" for good reasons of flight safety. He pointed to the common American rule of thumb that a military aviator should fly no fewer than 200 hours a year to maintain an acceptable level of proficiency. In sharp contrast, he said, "our pilots will soon not be able to fly at all due to the general shortage. . . . They are still managing to perform their assigned missions by using old reserves and resorting to barter deals. But what will it be like for Ukrainian airmen tomorrow?"

After the USSR's collapse, a debate unfolded in Ukraine over merging VPVO and VVS into a single service, with some espousing such a move on efficiency grounds and others countering that since air defense is paramount, all air operations should be subordinated to VPVO. The commander of Ukraine's VPVO, Lieutenant General Lopatin, advocated a go-slow approach. Not surprisingly, he also left little doubt about his commitment to a separate VPVO service, on the ground that "blind emulation of the structures of the West and the United States [where air defense is an organic component of the USAF] is not suitable for Ukraine and would be a major mistake."[57]

Not long afterward, a defense ministry collegium endorsed a proposal to create a new branch of the Ukrainian armed forces by merging VPVO and VVS into an integrated Air Defense Force.[58] This left unanswered the question of where Ukraine's air superiority and ground attack fighters would be lodged, to say nothing of the many other aircraft inherited from the USSR. The final resolution came in February 1993, with a reversal that brought Ukraine's VPVO and VVS assets into a Ukrainian Air Force under the command of Lieutenant

---

[56]Captain Vasily Verbitskii, "'Idle' Aircraft," *Narodnaia armiia* (Kiev), June 24, 1992.

[57]Lieutenant Colonel Nikolai Gorenko, "Responsible for the Sky: Today the Ukrainian VVS and VPVO Are Equally Responsible. However, Service Commanders Have Different Views on the Organization of Air Defense," *Narodnaia armiia* (Kiev), July 31, 1992.

[58]Vladimir Kaushanskii, "Soon There Will Be No Military Districts in Ukraine. Instead There Will Be Air Defense Forces," *Krasnaia zvezda*, September 5, 1992.

General Vladimir Antonets. Vasiliev and Lopatin were named deputy commanders in chief.[59]

The development of an independent air force by Belarus has been plagued by many of the same problems that have afflicted Russia and Ukraine. The commander of the Belarus Air Force, Lieutenant General Sergei Sedov, confirmed a serious shortage of fuel and spare parts, which had limited Belarus's pilots to no more than 40–45 flying hours a year (as opposed to what he said was a reasonable norm of 70–80 hours).[60] General Sedov voiced particular concern over the mass outflow of skilled personnel from his air force, in sharp contrast to Russia's pronounced surplus of such personnel. He added that if that process was not soon halted, "we will be unable to scrape together even one crew per aircraft. And in short order, we will lose our entire 'golden generation'—those who fly." He also noted that Belarus lacked a flight school for training replacement pilots and, unlike Russia, had inherited "a sufficient amount, even *too* much, of the most modern aviation equipment."

With the effective rout of the reform element led by President Stanislav Shushkevich in the January 1994 Belarus elections and its replacement by a more stolid leadership of new communists inclined toward reestablishing closer economic and military ties with Russia, the considerable front-line assets lost by the Soviet VVS to Belarus could eventually become resubordinated in some fashion to the Russian VVS.[61] Particularly now that Belarus has signed an inte-

---

[59]This mirrored similar mergers of air and air defense forces in several other former Warsaw Pact countries, notably Poland, Hungary, and the Czech Republic. For a good account of the formation of the Ukrainian VVS and the development of its doctrine, roles, and missions, see Ustina Markus, "Ukraine Restructures Its Air Forces: New Role, New Problems," *RFE/RL Research Report*, October 22, 1993, pp. 48–53.

[60]Interview by Colonel Valery Kovalev with Lieutenant General Sergei Sedov, "What Kind of an Air Force Will Belarus Have?" *Krasnaia zvezda*, November 17, 1992. The idea that 70–80 hours, low by Western standards, is an acceptable norm for maintaining operational proficiency is not wildly out of keeping with known Soviet practice. One Russian pilot told me that even in the best of times, a Soviet fighter pilot would not fly much more than 100–110 hours a year. Soviet pilots flew greater numbers of shorter-duration sorties, and they only logged actual flight time from takeoff to touchdown, with no increment for taxi time. Nevertheless, 100–110 hours a year is low compared with accepted Western practice.

[61]See Sonni Efron, "Pro-Communist Elected Head of State in Belarus," *Los Angeles Times*, January 29, 1994.

gration treaty with Russia, this is a possibility that bears careful watching.[62]

NEW PRIORITIES AND CONCERNS

General Deinekin faces an array of problems more challenging than those afflicting any other air force leader in the world today. Upon assuming his post as commander in chief, he inherited a near-total inversion of the priorities that typically concern a peacetime military aviation establishment. Matters like force modernization, training and tactics, and similar mission-related preoccupations took an immediate back seat to the more pressing demands of simply housing and caring for badly deprived personnel at a time when the VVS was reeling from the depredations of Soviet communism and the ensuing breakup of the USSR.

Among other vexations, General Deinekin continues today to be saddled with a severely curtailed procurement and operations budget, a fuel shortage of crisis dimensions, a bloated pilot-to-aircraft ratio further aggravating the insufficiency of available flying hours for Russian aircrews, widespread maintenance problems caused by a dearth of spare parts and the breakdown of military conscription, and an impacted and antiquated air traffic control system. Not surprisingly, the VVS has suffered a heightened aircraft accident rate as a direct or indirect result of these negative influences, along with a precipitous drop in the former prestige and respectability of air force service, with potentially grave implications for future officer recruitment.

The Collapse of State Financing for Defense

During the final days of its existence in late 1991, the Soviet defense ministry reported that outlays for weapons and associated procurement had fallen by 23 percent from the previous level in 1990. It anticipated that a comparable reduction would occur in 1992, meaning that defense production would be effectively halved from the base-

[62]See Michael Specter, "Belarus and Russia Form Union, Reuniting Two Former Soviet Lands," *New York Times*, March 24, 1996.

line 1989 level.[63] In the end, the VVS received only 15 percent of the allocations for research and development (R&D) and procurement that it was expecting in 1992. This forced it to buy equipment at the barest minimum level required to ensure that Russia's aircraft industry would not become completely moribund. Even such elementary provisions as flight suits and helmets are now in critically short supply.[64]

The head of the Central Finance Directorate of the CIS's joint command reaffirmed this trend toward diminished spending for equipment, offset by a commensurate growth in support for the quartering and welfare needs of Russian officers and their families. He estimated that some 70 percent of Russia's total defense expenditure for the first quarter of 1992 approved by the Supreme Soviet would be funneled into the social sector.[65] About 70 percent of the defense ministry's capital construction outlays in 1993 went to housing for military families.

Shortly before his appointment as First Deputy Minister of Defense, Andrei Kokoshin, then–deputy director of the USA and Canada Institute, predicted that Russia's defense industry would receive virtually *no* production orders in 1992, since all available funds had to be used to clothe and house military personnel.[66] Among Russia's airmen, an understandable concern emanating from this prompt reversal of spending priorities was that the VVS might eventually be gutted as a fighting force, much as occurred a generation earlier when, as General Malyukov put it, "many futures in aviation were destroyed at the end of the 1950s because of Khrushchev's excessive

---

[63]Radio Moscow domestic service, November 30, 1991.

[64]Lieutenant Colonel V. Rudenko, "Russia Scrambles the Tu-160," *Krasnaia zvezda*, August 4, 1992.

[65]Interview with Lieutenant General V. Vorobyev, Radio Moscow domestic service, February 14, 1992. Vorobyev frankly added that the sale of military hardware in the commercial market was "an uncharacteristic function" for the military, even though the need for self-financing had made it increasingly unavoidable.

[66]John Lloyd, "Sharp Cut in Soviet Defense Orders," *London Financial Times*, October 28, 1991.

fascination with missiles."[67] By late 1993, promised funding allotments from the Ministry of Finance had fallen so far behind, complained Kokoshin, that the defense ministry was a full trillion rubles in arrears to the defense industry for goods and services already delivered.[68] (That number increased to 2.2 trillion in 1994.) Indeed, added the chief of the ministry's Main Budget and Finance Directorate, the Russian civilian airline Aeroflot ceased honoring military transportation orders as of summer 1993 because of the enormous debt the defense ministry had piled up.[69]

The inertia of the old Soviet system, which routinely favored strategic missiles and armor, proved slow to die. In 1992, according to General Deinekin, aviation equipment accounted for only 12–15 percent of Russia's arms purchases, as contrasted to an asserted 25–30 percent in the United States. Since the USSR's collapse, the VVS has been forced to cancel any further purchases of the MiG-29. It has also had to defer development and production of several variants of the Su-27, which have been designated by the VVS as the intended mainstays of Russia's fighter inventory for at least the remainder of the 20th century.

Galloping inflation since President Yeltsin's elimination of state price controls in January 1992 has driven up the cost of modern aircraft astronomically. Research, development, test, and evaluation (RDT&E) on new platforms has largely been frozen, and the financing of several promising prototype programs has reportedly been halted (see Chapter Nine). In the United States, according to General Deinekin, the U.S. Air Force received 34–37 percent of all R&D funds budgeted in the 1991–1993 defense appropriation; the comparable figure for Russia was only around 15 percent. Warned then–Vice President Alexander Rutskoi, "if we ignore aviation today because we lack the money, it will take us 20–30 years to restore it."

---

[67]Interview by Dmitri Grinyuk and Piotr Butowski with Colonel General Anatoly Malyukov, "An Unusual Conversation at the Main Staff," *Krylia rodiny*, No. 11, November 1991.

[68]Nikolai Poroskov, "The Time for Stating Problems Is Past. Initiatives and Quests for Reserves Are What Is Needed," *Krasnaia zvezda*, December 9, 1993.

[69]Interview with Colonel General Vasily Vorobyev, "Simple Solutions Don't Exist in the Financial Sphere," *Krasnaia zvezda*, December 17, 1993.

As a stopgap measure, the VVS has sought nonbudgetary funding from domestic and foreign investors. To clear the way for this novel arrangement, a decree signed by President Yeltsin in February 1992 authorized the VVS to sell up to 1600 of its older aircraft to foreign buyers for hard currency. The idea was that any ensuing revenue would be channeled toward the production and operation of military aircraft in Russia, with the VVS acting as "chief client and guarantor." By one account, it was expected (quite unrealistically) that the VVS might collect up to $9 billion from such sales through the year 2000.[70] It was further hoped that such sales would, in addition to funding priority aviation programs, yield tax-exempt proceeds that might help the VVS build more housing and supplement officers' salaries.

Before the beginning of summer, however, there were reports of contraband weapons falling into the hands of rebels and gunmen in bordering former republics, ultimately leading defense minister Grachev to concede that criminal proceedings had been initiated against some suspected perpetrators.[71] Several months later, Grachev reported that as a result of the military's having been authorized to engage in commercial activities, "some servicemen failed the independent business test. . . . In our pseudomarket atmosphere, many went astray and couldn't resist abusing their official positions." A number of generals were fired outright as a result of such abuses. Others were reported to be under criminal investigation.[72] Retired VVS Major General Aleksandr Tsalko noted with disgust how this "approved" sale of military equipment to private buyers had become corrupted to a point where few of the proceeds actually reached the armed forces.[73]

---

[70]"The Air Force and Commerce," *Rossiiskaia gazeta*, March 26, 1992. According to a report in *Forbes*, a planned Russian "fire sale" at the Reus air base near Barcelona, Spain, was to have put MiG-29s on the block for as little as $100,000 apiece, and older fighters for as little as $20,000 cash—with the unstated proviso that any chance of securing replacement parts was virtually nil. The manufacturer's list price for the MiG-29 is around $20 million. See "Getting Tanked," *Washington Times*, April 13, 1992.

[71]Moscow television report, May 2, 1992.

[72]Account of a speech by General Grachev at the Seventh Congress of Russian People's Deputies, *Krasnaia zvezda*, December 8, 1992.

[73]Moscow television service, December 12, 1992.

Understandably, General Deinekin was reluctant to take on the embarrassing question of alleged corruption by unnamed VVS colleagues in the commercial sale of military equipment. He would only concede that in July 1992, President Yeltsin's decree authorizing aircraft sales had been rescinded as a result of "a series of palace intrigues." He would neither confirm nor deny press allegations of high corruption on a vast scale in the Ministry of Foreign Economic Relations with respect to the sale of Russian arms and military hardware.

A Growing Pilot Surplus

The VVS's crew ratio, or the number of line pilots per operational aircraft, has more than doubled since the collapse of the USSR. General Deinekin stated in early 1992 that it had risen to three pilots for each flyable aircraft because of force reductions and accelerated unit withdrawals from Eastern Europe and the former Baltic republics.[74] He later remarked that in some units the crew ratio had become as severe as *five* pilots per aircraft.[75]

This pilot glut is especially concentrated in fighter and ground attack units. According to the head of the VVS's Training and Assignment Directorate, Major General Osipenko, voluntary withdrawals of rated VVS personnel from active flight status at the end of 1992 posed no threat to the VVS's projected pilot needs. On the contrary, there was a requirement to *reduce* overall pilot strength by at least 25 percent merely to stay abreast of continuing unit deactivations and force reductions.[76] That requirement has no doubt increased in subsequent years.

One approach toward grappling with this problem that has been aired at VVS headquarters has been to encourage fighter pilots who wish to remain on flight status to volunteer for other aviation

[74]Interview with Colonel General Deinekin, "A Country Without Wings? No," *Trud*, March 10, 1992.

[75]Lieutenant Colonel A. Vetakh, "Waiting for an Aircraft: Combat Pilots Await Their Turn," *Krasnaia zvezda*, May 15, 1992.

[76]Interview with Lieutenant General Aleksandr N. Osipenko, "We Will Have Enough People Who Desire and Are Trained to Fly," *Krasnaia zvezda*, December 18, 1992.

branches or to accept navigator assignments. As a triage technique for managing its aircrew reduction plans, the VVS is treating those pilots who have served three to four years in a given assignment as a "reserve" pool for potential selection to higher positions. Others with five or more years in the same posting who are considered poor prospects for promotion will most likely end up being released into the reserve.

Because of the pronounced shortage of cockpits, General Osipenko conceded that there was "no chance" that every 1993 flight school graduate could be guaranteed a flying assignment. A sizable number of freshly minted pilots were banked in temporary jobs on headquarters staffs, in command posts, or as technicians, with the understanding that they would eventually be moved on to a flying assignment once one became available. As a matter of policy, the VVS will not force a flight school graduate to serve in a nonflying assignment outside his specialty. Accordingly, those newly winged pilots who declined banked assignment options were graduated with the rank of lieutenant and summarily discharged into the reserve.[77]

The VVS is striving to reduce its crew ratio to a stabilized norm of three pilots for every two aircraft. In the meantime, the pilot surplus has imposed a considerable burden on day-to-day training in operational squadrons. A case in point is the senior lieutenant who described taxiing out for a long-awaited flight to a weapons range to maintain his mission currency, only to experience an avionics system failure immediately prior to takeoff. The result was a noneffective sortie. The lieutenant later remarked: "The aircraft situation here is really like a free-for-all. You should see how emotions flare up when we are preparing our little 'plan.' Each pilot and flight commander thinks that his problems are the most important. What happens is that everyone keeps pulling the blanket over to his side. . . . All these gyrations are prompted by the growing number of pilots arriving from VVS units undergoing reductions and, for other reasons, from various areas of the former USSR. But the aircraft pool remains the same."[78]

---

[77]More than 100 new cadets were enrolled at the flight school at Tambov in 1992. Among its graduates the same year, 40 were immediately released into the reserve.

[78]Quoted in Vetakh, "Waiting for an Aircraft," *Krasnaia zvezda*, May 15, 1992.

The Crisis in Flying Hours

Following President Yeltsin's lifting of price controls in January 1992, fuel costs escalated by 2000 percent during the remainder of that year.[79] The first deputy head of the defense ministry's Main Budget and Finance Directorate reported that because of reduced appropriations for fuel, pilots were typically getting less than a third of their annual flying norm.[80] One article noted that because of an excess of flight personnel and the severe shortage of petroleum, oil, and lubricants (POL) and spares, the average flight time accrued by Russian fighter pilots during the first ten months of 1991 was less than 40 hours, and this, not surprisingly, in regiments that were experiencing the most accidents.

With no realistic prospect for operational aircrews to meet even their minimal mission currency requirements, the author proposed that the VVS should suspend its published training norms, at least for the time being, since their persistence on the books merely encouraged unit commanders to engage in dishonest reporting. He further remarked that many VVS pilots are accidents waiting to happen: "Pilots are still languishing while awaiting their chance to fly. Once they do get airborne, their commanders worry—will their pilot, having such limited proficiency, make it back to his base in one piece?"[81]

Even before the USSR's collapse, a senior pilot wrote of casually perusing several squadron-mates' flight logs and noting that in one month, one pilot had flown ten sorties for seven hours, in the next month seven sorties (six day and one night) for four hours, and the next month only two sorties for barely more than an hour. This partly reflected, he said, the effects of a self-inflicted VVS "prohibition mania," whereby fighter units are forced to suspend *all* flight activity in the wake of an accident until the causes are determined—even if the accident occurred in a helicopter or transport squadron! The net result, he added, is merely to aggravate the existing safety situation (see below), since noncurrent pilots are more

---

[79] *Aviation Week and Space Technology*, January 11, 1993.

[80] Interview with Major General Vasily Kuznetsov, "Inflation Devours the Military Budget Before It Can Be Approved," *Krasnaia zvezda*, October 15, 1992.

[81] Colonel Yu. Timchenko, Colonel S. Shumilo, and Lieutenant Colonel S. Bolotin, "Is the General Always Right?" *Aviatsiia i kosmonavtika*, No. 2, February 1992, pp. 10–11.

accident-prone than proficient ones. Worse yet was the reflexive tendency of many unit commanders to impose a determined "back to basics" approach in the wake of an accident. Because it erred in imposing blanket bans on experienced and junior pilots indiscriminately, even the most proficient pilots would "slip willy-nilly back to the level of average ones."[82]

General Deinekin confirmed in early 1993 that largely because of the fuel shortage, VVS fighter pilots were averaging 40 flying hours a year, bomber pilots 80 hours, and VTA pilots 150 hours (the differences reflecting variations in mission type, with LRA and transport crews flying fewer sorties of longer duration).[83] General Deinekin further reported that the VVS at the time had roughly two assigned pilots for each single-seat aircraft, since several thousand fighter pilots stationed in the former republics had returned home following the collapse of the union.[84] In most cases, the fuel shortage has required regimental commanders to preclude their staff officers from flying altogether, so as to assure the most rational distribution of their meager fuel allotments to their neediest line pilots.

Conditions were scarcely better in 1992 with those VVS fighter units that were awaiting final withdrawal from eastern Germany. The air commander for the Western Group of Forces (WGF), Lieutenant General Tarasenko, remarked that to give each pilot an equal chance to fly in such circumstances would be tantamount to providing an opportunity to no one, since "letting everyone fly, but no more than once or twice a month, would mean taking everyone to the brink of losing his professional skills."[85]

---

[82]Guards Major A. Gornov, "Prohibition Mania," *Aviatsiia i kosmonavtika*, No. 5, May 1991, p. 30.

[83]General Deinekin has stated that in the best of times he used to fly 500 hours a year when he was a line pilot in LRA. That would be a dream for any LRA pilot today. After I flew a MiG-23UB with the Mikoyan Design Bureau at the Ramenskoye Flight Test Center in August 1993, my instructor pilot, Colonel Vladimir Gorbunov, told me that there was no shortage of jet fuel per se. The problem, he said, was simply a shortage of money to pay for it.

[84]Report by ITAR-TASS correspondent Vadim Byrkin, February 16, 1993.

[85]Interview with Lieutenant General A. Tarasenko, "In a Holding Pattern: Comments on Problems of the 16th Air Army," *Krasnaia zvezda*, March 27, 1992.

While awaiting the withdrawal of one WGF regiment from Germany in 1993, a deputy commander reported that his unit was being allotted only 100 tons of jet fuel a month, the amount previously apportioned for a single *day's* training schedule. With approximately three tons of fuel consumed on a typical 30-minute fighter training sortie, the result, said Colonel Novikov, was predictable: "Figure it out yourself. You come up with 33 flights for sixty aircrews." A regimental commander, Colonel Borisyuk, added that "it pains the soul to think of the fate of our pilots," since "in every civilized nation aircrews are regarded as a real treasure." Borisyuk went on to note that "a true pilot will never, of course, actually forget how to fly." However, he stressed, a minimum of three flights a month was "the lower threshold that must not be crossed." Now down to 25 hours a year, Russia's line fighter pilots are flying just enough to be dangerous to themselves and others.

A later account of WGF training indicated similar currency and proficiency concerns as forward-based VVS units approached the midpoint of their three-year phased withdrawal from former East German territory. The deputy commander of Russian forces in Germany, Major General Nikolai Seliverstov, reported that available flight time for WGF pilots had been cut back to the bone and that missions "in zone" had grown progressively more rudimentary as a result of the disappearance of any operational purpose behind the lingering Russian presence in Germany. A Third-Class pilot stationed at Finow remarked that he anticipated flying no more than 40 to 50 hours in 1993.[86]

The fuel shortage has affected more than Russia's fighter pilots. Colonel General Igor Kalugin, commander in chief of Long-Range Aviation, reported that flying time for his bomber crews had also fallen to crisis levels because of fuel limitations and the declining service life of many LRA aircraft. General Kalugin added that he had been forced to limit his bomber crews to flying combat aircraft only

---

[86]Jeffrey M. Lenorovitz, "Reduced Russian Operations in Germany Eroding Pilot Proficiency, Readiness," *Aviation Week and Space Technology*, June 1, 1992.

along fixed navigational routes and to using a Tu-134 jet transport as a bomber surrogate for most routine currency training.[87]

Colonel General Anatoly Borsuk, then-head of the VVS's Combat Training Directorate, noted that engine and spares availability had fallen off dramatically over the preceding five years. He lamented the poor reliability both of new equipment and of older aircraft that had long been in squadron service. He further complained of a shortage of flight simulators, a growing dearth of engineering and technical support personnel, and an excess of aircrews resulting from the unit shutdowns within Russia and the continuing withdrawal of VVS regiments from Eastern Europe mandated by the Conventional Forces in Europe (CFE) Treaty. In 1992, said Borsuk, a VVS fighter pilot's annual flying allotment was 2.5 times less than the ideally desired amount, and three to four times less than that said to be made available for American military pilots. Some VVS pilots, he said, go without flying for two and a half to three months. By that time, even in LRA, they are no longer current and must be retrained.[88]

In a revealing snapshot of where things stood a year later, a military reporter provided an arresting account of VVS flight activity during a typical 24-hour day in the fall of 1993. A conversation with Major General Aleksandr Slukhai, senior duty officer in the central command post at VVS headquarters on the day in question, indicated that VVS flight schools and fighter aviation recorded 845 sorties that day for a total of 459 flying hours, with LRA registering 183 sorties for 115 hours, and VTA logging 117 sorties at training centers for 58 hours in the air. The total came to slightly more than 1000 VVS flights, for an average sortie length (*including* in LRA and VTA) of around a half-hour each. The reporter tried hard to put the best possible spin on these figures: "There is no basis for the idle conjectures of certain mass media that the VVS has neglected combat training. . . . A total of 1145 training flights in a 24-hour period—is that not combat training?" The bitter truth, however, was laid bare in General Slukhai's more disquieting observation: "Some days the

---

[87]Interview with General Kalugin, "In the First Strategic . . .," *Aviatsiia i kosmonavtika*, No. 10, October 1991, pp. 2–3.

[88]Interview with Colonel Aleksandr Fomin, an LRA regimental commander, "A Regiment Commanded from Three Countries At Once—On Top of This, the Commander Is Threatened With Court Action," *Krasnaia zvezda*, April 18, 1992.

flying time for the entire VVS adds up to the number of hours the regiment I previously commanded would have flown in a 24-hour period."[89] In a telling contrast, General Deinekin reported earlier that on a typical flying day in August of the preceding year, the VVS had registered 6798 sorties.[90]

It is quite possible that the VVS will encounter even worse difficulties in resolving its fuel crisis on a nationwide basis, at least in the near term, in light of the increased regionalization of fuel development and the expanded growth of market relations throughout Russia. A joint stock company called the Russian Innovational Fuel and Energy Company (RITEK) was established in March 1992 expressly to deal with the aviation fuel crisis. However, its immediate beneficiaries were in Western Siberia. Only after a number of refineries were put into operation elsewhere was it expected that the fuel crisis could be alleviated in other regions. From the meager details reported, it was not clear whether the main beneficiaries of this development would be military or civilian.[91] It is noteworthy that growing numbers of VVS officers have come to admit freely, albeit in private conversations, that organized crime in Russia now wields a virtual hammerlock on the VVS's suppliers, including those in the fuel industry.

Maintenance Shortcomings

Much like the USAF, the VVS operates a three-tiered aircraft servicing system that includes line, base-level, and depot maintenance.[92]

---

[89]Quoted in Aleksandr Manushkin, "Things Are Never Boring for the Duty General," *Krasnaia zvezda*, October 23, 1993.

[90]Weather may have been an extenuating factor here, considering that the sortie number provided by General Slukhai was recorded in October, while that given by General Deinekin was recorded in August, when flying conditions in Russia are best. Nevertheless, the difference speaks for itself. General Deinekin further noted that on that same flying day in August 1992, 980 sorties were reported by PVO, 409 by the Strategic Rocket Forces, and 432 by the Navy. Interview with Yelena Agapova, "A Russia Without Wings Is Not Russia. It Does and Will Have Them," *Krasnaia zvezda*, August 15, 1992.

[91]See B. Sidorenko, "RITEK's First Steps: A New Company Will Have an Influence in Overcoming the Aviation Fuel Crisis," *Vozdushniy transport*, No. 42, October 1992.

[92]For a good background treatment of VVS maintenance practices during an earlier period of the cold war, see Andris Trapans, *Organizational Maintenance in the Soviet Air Force*, Santa Monica, California, RAND, RM-4382-PR, January 1965.

Base-level aircraft maintenance (the equivalent of USAF intermediate maintenance) is performed by the regimental Technical Maintenance Unit (*Tekhnicheskaia ekspluatatsionnaia chast*, or TECh), which additionally conducts scheduled inspections every 600 and 1200 hours.

Problems with quality control have long plagued VVS maintenance. These problems have become considerably more severe in the period since the USSR's collapse. General Shaposhnikov reported that the VVS had managed to sustain a 90-percent aircraft in-commission rate during the penultimate year of the USSR.[93] Just a year later, however, to cite a representative example, one regiment reported that only 25–30 percent of its Tu-22M Backfire bombers were serviceable, with the rest out of commission because of engine or other problems. The general in charge of acquisition in the defense ministry attributed this to a lag in manufacturing technology and poor quality control at the production line, along with a shortage of modern production tooling and poor discipline at aircraft and engine factories.[94] That, however, was only a part of the explanation.

VVS maintenance manning in 1991 was a third below assigned strength, with only half the needed number of replacement personnel being provided by the various training schools.[95] That was in considerable part a consequence of post-Soviet Russia's failed conscription system. Largely owing to the generous student deferment provisions (since withdrawn) approved by the Russian parliament and the refusal of most other draft-age males to honor their call-up notices, defense minister Grachev predicted that only 26 percent of the anticipated number of draftees nationwide would report for induction in 1993.[96] In the important Moscow Military District, the expected number was as low as 3 percent.

---

[93]Colonel General Yevgeny I. Shaposhnikov, "Results and Prospects," *Aviatsiia i kosmonavtika*, No. 1, January 1991, p. 2.

[94]Interview with General Vitaly M. Shabanov, "Military Parity Is Not a Game of Give-Away," *Kommunist vooruzhenykh sil*, No. 13, June 1991, p. 12.

[95]Guards Major V. Fomin, "Seconds Equal to Years," *Aviatsiia i kosmonavtika*, No. 3, March 1991, pp. 6–7.

[96]Interview in *Izvestiia*, June 2, 1992.

As a result, noncommissioned manning has fallen to the 50-percent level or below in many VVS and VPVO units. "This is an alarming indicator," said VPVO commander Prudnikov, "because it was always felt that a unit was not operationally ready if it dropped below 70 percent. We have now crossed that line."[97] Even before the August 1991 coup, the VVS's deputy commander for logistics complained that maintenance manning remained well below its mandated strength. He said that as a result, the VVS was able to provide only some 60 percent of its required rear-service support for training and readiness.[98]

In the face of this manpower shortage, regimental commanders have frequently cut the number of maintenance personnel assigned to flying squadrons arbitrarily, without first weighing the likely effects on maintenance delivery.[99] The shortage of skilled manpower has further obliged unit-level maintenance sections to assign barely trained conscripts to serve as aircraft mechanics.[100] For good reason, VVS maintenance professionals complain that "the soldier in aviation is of little help to the officer technician." The VVS's maintenance schools for conscripts yield poorly trained graduates, few of whom develop any significant skills by the end of their 24-month service period. Most end up merely performing guard duty and attending to housekeeping chores. Alexander Zuyev recalled that many of these conscript mechanics could barely read Russian and had to be instructed using the same rote techniques one would use with a child. For that reason, fighter regiments rely on a small core of trained maintenance officers, supported by warrant officers who su-

---

[97]Interview by Colonel A. Belousov and Major A. Ivanov with Colonel General Prudnikov, "A Unified System Is Needed," *Krasnaia zvezda*, April 11, 1992.

[98]Operational units in all services in 1992 were reportedly manned on average at only around 60 percent of their assigned strength. Statement by the chief of the General Staff, Colonel General Mikhail Kolesnikov, cited in Viktor Litovkin, "The Army Pins Hopes on the Help of Legislators . . . and Women," *Izvestiia*, October 17, 1992. During the first half of 1992, nearly 70 percent of Russian youths who were eligible for service dodged the draft. Sergei Ostanin, ITAR-TASS, October 15, 1992.

[99]Major N. Barabanov, "Flying Safety Suffers," *Krasnaia zvezda*, June 20, 1991.

[100]Lieutenant General G. Shinkarenko, "I Don't Want to Be an Instructor," *Krasnaia zvezda*, March 12, 1988.

pervise the conscripts and bend every effort to prevent them from "destroying the aircraft."[101]

As a rule, functional check flights (FCFs) following routine maintenance are not performed because of the scarcity of fuel. This naturally engenders an indifferent attitude on the part of maintenance toward aircrew write-ups. That, in turn, lowers aircrew confidence in their equipment. A TECh officer complained that maintenance understaffing was causing minor write-ups to go unattended until the next scheduled intermediate maintenance, in the blind hope that the TECh might discover and fix the problems.

VVS fighter squadrons also do not routinely require formal postflight maintenance debriefs by the pilot. Nor, apparently, do they maintain detailed logs for recording and tracking avionics anomalies. Among suggested interim fixes for these problems have been the use of flight recorders to monitor the performance of the fire control system to aid in postflight troubleshooting, and the conduct of postmaintenance system checks during scheduled training sorties (in effect, an FCF on the run), on the premise that any such checks, however haphazard, are better than none.[102]

With the introduction of fourth-generation fighters like the MiG-29 and Su-27 into its inventory, the VVS has encountered recurrent problems with fault isolation in avionic systems capable of multiple failure modes, much as the USAF experienced with the F-15 during its first years of operational service. "At the outset," complained one Russian officer in a refrain familiar to USAF avionics technicians, "everything is fine. Then an anomaly occurs. By the end of the flight, everything is back to normal again. It is extremely difficult to detect such a stray defect on the ground."[103]

Adding to this problem considerably, much of the VVS's maintenance support equipment is rudimentary. Setting up a diagnostic

---

[101]Zuyev, *Fulcrum*, p. 136.

[102]Not surprisingly, pilots routinely complain about maintenance. One commented sarcastically that "we have two probable enemies, NATO and ORATO" (the latter being the Russian acronym for the VVS's auxiliary airfield technical support unit). Major N. Chebotarev, "For the Planes to Fly . . .," *Aviatsiia i kosmonavtika*, No. 1, January 1993, p. 14.

[103]Ibid.

system at the regiment level often calls for artful and aggressive scrounging skills, plus complete reliance on the unit's own resources. Because of the widespread unavailability of computers, it is particularly difficult to monitor avionics status and predict failures. Accordingly, some navigation and weapons delivery modes are not used at all because of their poor accuracy and reliability.

Cannibalization of parts from some aircraft to keep others flying has become common in many fighter units, even though it is in direct violation of safety rules. Such reliance on so-called "donor aircraft" (a polite term for hangar queens) was bound to occur sooner or later as a result of the steadily declining availability of assemblies and spare parts. The impetus behind this flouting of published rules and good judgment has been to keep the greatest possible number of aircraft flyable at any cost, since flying hours are meted out according to the number of serviceable aircraft in a given unit. Even *with* cannibalization, some units have lost considerable flight time as a result of delays in the delivery of tires, POL, and other consumables. Wrote one officer, "we will find a way out of this situation by hook or by crook, including by cannibalizing aircraft. But what about tomorrow?"[104]

A shift to contract maintenance is widely portrayed as the only workable solution over the long haul, since such an arrangement would "nurture a work environment conducive to the development of an incentive in each technician to become a bona fide professional." The situation has been further aggravated by Spartan and even forbidding work conditions in many cases. Complained one navigator assigned to the Transbaikal Military District: "We are flying on scrap metal. The equipment is old. There are virtually no spares, nor any facilities for repair. People in the squadron often say: 'Jet fuel and blood are mixed together with us.' Imagine forty degrees below freezing. An exposed flight line. People working with bare hands.

---

[104]Lieutenant Colonel A. Vetakh, "'Donor' Aircraft," *Krasnaia zvezda*, April 15, 1992. Colonel General Kalugin singled out the Tu-160 as a maintenance nightmare because of its unusual complexity. The aircraft requires a dedicated air base equipped with special ground-support equipment, high-pressure hydraulics, and an extensive supply train. Kalugin complained that its design is still being debugged and that a more user-friendly and less expensive bomber is needed.

Their fingers split. So you have blood and kerosene. Of course we are unhappy. But we serve. Someone has to defend the country."[105]

Safety and the Accident Situation

All of this has had a predictable impact on flying safety in the VVS. Shortly after becoming commander in chief, then-Colonel General Shaposhnikov admitted to "several dozen" aircraft accidents in 1990, with 60 percent involving equipment in good working order.[106] Shaposhnikov conceded that any improvement of the situation would require the prior solution of "a whole host of problems associated with VVS life and activity."

The VVS showed an increase in the number of major mishaps in 1992, the first year following the collapse of the USSR. However, the head of the Flight Safety Service (*Sluzhba bezopasnosti poletov*, or SBP), Major General Alekseyev, insisted that this did not constitute grounds for immediate alarm, since "even in the most favorable years for the country, the state of safety was only a bit better."[107] Alekseyev confirmed that for decades the proportion of flight mishaps due to pilot error had exceeded 60 percent. The big difference by that time—1993—was that recurrent failures to implement effective preventive measures had been amplified by new plagues against the health of the flying community, including a lack of adequate fuel allocations, flight simulators, support equipment, and pay. These problems have only worsened in the intervening three years.

A report in June 1992 declared that "the aircraft accident rate is threatening to shift from isolated instances to a landslide."[108] It noted that there were 26 major mishaps in VVS operating units in 1991, with eight recorded during the first three months of 1992 alone.

---

[105] Captain Vladimir Pasternak, quoted by Colonel Viktor Baranets, "Word to Rally," *Pravda*, January 17, 1992.

[106] Colonel General Yevgeny I. Shaposhnikov, "Results and Prospects," *Aviatsiia i kosmonavtika*, No. 1, January 1991, p. 2.

[107] Interview with Major General Aleksei I. Alekseyev, "The Formula for Reliability: Man—Aircraft—Environment," *Krasnaia zvezda*, January 13, 1993.

[108] Comment on a letter to the editor from a Zhukovskii Air Force Engineering Academy student by Colonel A. Andryushkov, "We Are Flying Less and Less. We Are Crashing More Often. Will Russia Become Wingless?" *Krasnaia zvezda*, June 19, 1992.

The article added that in some regiments, pilots were not even getting a minimal allocation of 40 flying hours a year, and that it was precisely in those units where the accident rate was most disturbingly on the rise. It implored the VVS to take a hard look at proven foreign aviation safety practices in search of a better way to ramp down the incidence of flight mishaps. It also stated that in 1968 the Soviet Air Force roughly matched the USAF in the number of accidents per 100,000 hours, whereas today the VVS exceeded the USAF's rate by a factor of two, even with "many times" fewer flying hours.

More recently, General Deinekin disclosed that the VVS was suffering some 50 fatalities and upward of 100 aircraft losses a year in routine training accidents.[109] The majority of those as well were caused not by equipment failure but by pilot error, with most occurring to First-Class pilots.[110] Deinekin went on to report that there was an increase in the number of aviation-related fatalities in 1992 because of several mishaps involving transports, and that the flight safety environment had worsened notably. One of the chief reasons, he said, was the collapse of stable financing for fuel purchases. General Deinekin reported that the VVS was granted only half the fuel allotment in 1992 that it received in 1991, and that because of irregular deliveries the supply for that year was effectively down 20 percent more.

Several years earlier, Marshal Kirsanov, then–deputy commander of the Soviet Air Force, had faulted the VVS's tendency to focus on ferreting out the most proximate cause of an accident as a basis for parceling out blame, without probing deeper for associated causes that might have been more pertinent as root explanations for the accident.[111] He also singled out the burdensome collateral duties

---

[109]Interview with Colonel General Deinekin, "The Man Sitting In Moscow Is Going Into a Spin Again . . .," *Komsomolskaia pravda*, March 17, 1993.

[110]Interview by Yelena Agapova with Colonel General Deinekin, "The Skies Are the Same—the Concerns Different," *Krasnaia zvezda*, October 26, 1991.

[111]Marshal of Aviation P. Kirsanov, "There Are No Minor Details on This Question," *Krasnaia zvezda*, April 22, 1988. Confirming that the VVS has had its share of avoidable accidents, Marshal Kirsanov gave as examples a pilot who flew too slow and spun his aircraft during an intercept; pilots who flew through their own bomb fragmentation patterns during live weapons drops on tactical ranges; maintenance personnel who forgot to properly fuel an aircraft before takeoff; and controllers who cleared aircraft for takeoff with another aircraft obstructing the runway.

levied on squadron pilots as a hindrance to the maintenance of an adequate level of aircrew proficiency.[112] Such distracting claims on a young pilot's time, he said, can lead not only to catastrophic accidents, but worse to a calculated compromise of integrity, as reflected in such dishonest actions as "redoing flight planning tables or falsifying write-ups on the plans and logs," a practice commonly disparaged throughout the USAF as "pencil-whipping." He said that commanders who tolerate, or themselves indulge in, such shortcuts instill in their subordinates a habit of "merely processing the paperwork properly and getting away with a violation."[113]

In a similar expression of high-level candor, the respected commander of VPVO fighter aviation, Colonel General Vladimir Andreyev, conceded that Soviet efforts to grapple with the safety problem had consisted mainly of idle talk for thirty years. He added that the problem will never be fixed merely by "words and threatening directives." To illustrate the extent to which the system had lost sight of the big picture, he recalled once asking his subordinates at VPVO headquarters, just as an experiment, to come up with a list of questions to which a MiG-31 pilot needed to know the answers. "And they produced . . . a 900-page book! Nobody can assimilate that much detail," he responded. "And the pilot does not need it! You need to give a pilot the minimal amount of knowledge necessary to allow him to fly his aircraft responsibly. And then—let him improve himself. With that, there is no limit."

Continuing, said General Andreyev: "If a pilot has mastered the minimal amount of knowledge, we won't torture him with more the-

---

[112]These included 24-hour watch duties, including supervising conscripts, inspecting barracks and mess halls, and generally enforcing discipline. John Barron, *MiG Pilot: The Final Escape of Lieutenant Belenko*, New York, McGraw-Hill Book Company, 1980, p. 96.

[113]Similar criticism was offered by a retired major general: "In investigating an accident, efforts must be aimed not at looking for a culprit or at what was done wrong by those who got into trouble, but at what is wrong with the management system, with organization, with the technological process, and with production activity." This officer charged that "the command-pressure method of leadership continues to shake loose the foundations of air unit flight safety and combat training." He attributed this to "inertia in thinking, adherence to stereotypes, and, most important, the fear of 'what if something happens?'" Major General of Aviation (Ret.) A. Bystrov, "The Concept of Preventing Mishaps: Theory and Practice," *Aviatsiia i kosmonavtika*, No. 8, August 1990, pp. 16–18.

ory. We'll clear him for three months and let him fly with God. Later, we'll check to see whether he has grown or stagnated. As a rule, self-improvement occurs." However, added Andreyev, for any such system to work, a culture change would be needed in the flying community. "If we wish to be called professionals," he said, "we in the military first of all must put a decisive end to formalism, bureaucratism, dishonesty, and cooking the figures in every conceivable way."[114]

In a related argument, a senior navigator chided both the Combat Training Directorate and the Flight Safety Service, whose leaders he portrayed as sometimes being "more zealous about faulting each other than about finding constructive solutions." The price of concentrating solely on apportioning blame rather than on understanding the cause of accidents and developing appropriate measures for preventing recurrences, he added, is that it invariably results in "the papering over of dangerous situations that will surely recur."[115]

Arbitrariness in accident investigations, however, remains a continuing problem in Russian military aviation. A case in point followed a MiG-31 mishap in which the aircraft experienced violent uncommanded pitch and roll oscillations shortly after takeoff. The crew finally ejected successfully only moments before the aircraft struck the ground. The ensuing debrief of the pilot and eyewitness accounts of the mishap both confirmed a mechanical failure on the aircraft. Yet the accident board reported the cause to have been the pilot's having attempted a mission "beyond his capabilities." This finding was rejected by all aircrews in the parent regiment as patently bogus. Yet

---

[114]Interview with then–Lieutenant General Vladimir I. Andreyev, "We Need to Know the Threat by Sight," *Krasnaia zvezda*, November 22. By "threat," General Andreyev had in mind not the United States or NATO, but situations that lead to flying accidents: "The word 'safety' (in Russian, *bezopasnost*, or 'absence of danger'), both in its derivation and in life, has as its root 'danger.' In order not to end up at risk, you need to know risk well, by sight, as they say, and to know how to act not to end up in danger and to know how to deal with it. . . . Efforts to introduce this approach to training pilots have not gone very smoothly. Rumors have circulated that VPVO pilots are being made to 'study some sort of dangers.' Our critics have not understood that in imparting knowledge of danger, we are saying what matters most, namely, that a pilot can recognize and emerge successfully from an encounter with a nonstandard situation."

[115]Colonel V. Dudin, "A Pilot Can Encounter an Emergency: Is He Ready to Master It?" *Krasnaia zvezda*, September 7, 1991.

the only "preventive measures" implemented in the wake of the investigation were to relieve the two crew members of their squadron duties and to ground the pilot.[116]

A related example of how the fuel crisis and resultant reduced flying hours have left their mark on the accident rate was a fatal mishap involving a Su-24 following a failure of the left afterburner to light during takeoff roll. The resultant asymmetric thrust caused by the loss of power on the affected engine produced an uncommanded roll into the ground immediately following a takeoff that should have been aborted. An assessment of the accident afterwards concluded that the lapsed pilot proficiency that underlay this particular mishap represented "the chief risk factor in conditions of a sharp reduction in flying hours."[117]

Problems with Air Traffic Control

The VVS inherited a Byzantine air traffic control (ATC) and flight clearance system from the former Soviet Union. To secure approval for a scheduled flight from one military airfield to another in a different center's jurisdiction, a pilot must submit an airspace reservation request to the controller at his home airfield two hours before his planned departure. After that, the request moves in sequence through the home unit's command post to the military sector of the regional civilian ATC center, the zonal center, the ATC center at the destination airfield, and finally to the individual controller at the destination airfield. Only then, with the home regimental commander's approval, can the flight be cleared to depart.[118]

Controllers with transit approval authority have little incentive to facilitate the movement of air traffic through their jurisdictions.

---

[116]Colonel N. Ryabnikov, Lieutenant Colonel M. Subbotin, and Lieutenant Colonel S. Bolotin, "So Just What Did Happen?" *Aviatsiia i kosmonavtika*, No. 7, July 1992, p. 13.

[117]Colonel Yu. Timchenko, "Thirty Seconds During a Takeoff . . .," *Aviatsiia i kosmonavtika*, No. 5, May 1993, p. 10. A good review of recent debate over accident investigation and flight safety management in the VVS is provided in Dennis Marshall-Hasdell, *The Reform of Flight Safety in the Soviet Air Force*, Soviet Studies Research Center, Royal Military Academy Sandhurst, Camberley, England, February 1993.

[118]Lieutenant Colonel V. Skurikhin, "Cleared for Departure," *Aviatsiia i kosmonavtika*, Nos. 3–4, March–April 1992, pp. 5–6.

"Quite the contrary," complained one pilot, "the fewer that are flying, the less the resultant hassle. As they say, God forbid that anything should happen." Pilots joke about one controller said to have never authorized a single aircraft to pass through his assigned airspace throughout his entire career as a duty officer! Seasoned pilots do not even bother wasting their time submitting a cross-country flight request during the last hour before a controller shift change, during mealtime, or any time on the day before a holiday.[119]

This particular pilot recounted a nightmare experience during which he was once forced to lay over at a civilian airfield for two days in the course of a cross-country flight. First, he could not get fuel. After he finally scrounged the fuel, the weather deteriorated. "But typically," he said, "for some reason it only deteriorated for us military fliers. Cross-country civilian crews at that point were still not being delayed. . . . I'm surprised the VVS leadership hasn't yet figured out why aircrews from various ministries and agencies seek in every conceivable manner to avoid landing at military airfields." He added: "The time has come to review the current structure of air traffic control points, eliminate redundant echelons, and make the 'unified' ATC system truly unified." Since the main responsibility for coordinating flights, including military flights, resides within the civilian component of the ATC system, he added, it makes no sense to retain military ATC centers other than where they are needed because of unusually dense local military traffic. He suggested that this would minimize friction between the military and civilian components of the ATC system. He also contrasted the hidebound Russian system with that of the United States, "where they only control rather than command air traffic" and where clearances can routinely be processed in half an hour.

Post-Soviet Russia is just now beginning to modernize this encrusted system. Even before the breakup of the USSR, there was an acknowledged problem of artificial jurisdictional barriers. The commander

---

[119]Characteristically, graft was a common lubricant of the Soviet system. The head of the Interdepartmental Commission of the unified ATC system, Major General Boris Kushneruk, complained how one VTA crew on a scheduled military mission was delivering refugees from hot spots and upon arrival was shaken down for "a tidy sum for landing support. Where is the crew supposed to get the money?" Interview by Vitaly Moroz, "Can the Skies Be Privatized? Military Aviators Are Convinced They Cannot," *Krasnaia zvezda*, December 22, 1992.

of VPVO communications reported that the VVS, Navy, and Ground Forces each monitored their own portion of Soviet airspace alongside the civilian ATC system. VPVO also maintained its own radar surveillance, controlling up to 12,000 flights a day through a mere 150-km strip of airspace. A move toward a reconfigured Unified Air Traffic Control System (*Yedinaia sistema upravleniia vozdushnogo dvizheniia*, or YeS UVD) was finally prompted by the realization that it made no sense for controllers of these overlapping jurisdictions to be sitting often literally side by side, yet receiving only that information pertinent to their own operational concerns.[120] A portion of the new system was slated to be tested in 1995, but a lack of funding has most likely brought its implementation to a standstill.

The Declining Quality of VVS Life

Three years before the collapse of Soviet communism, a former VVS officer (now executive director of the influential Russian Council on Foreign and Defense Policy) offered a rare glimpse behind the myth of privileged life in the officer corps when he suggested that if unit commanders could only gain state permission to use the government funds that they had managed to save through frugal spending to build housing, kindergartens, and other social facilities for VVS personnel, they could inspire a major savings campaign and elicit widespread support from below.[121]

In a similar vein, six months before the August 1991 coup, then–VVS commander Shaposhnikov attacked the inadequate provision of housing and social amenities for the families of VVS officers. Some of this he blamed on the return of Soviet units from Eastern Europe at an unexpectedly rapid rate. However, Shaposhnikov complained that responsibility for the VVS's housing conundrum lay primarily with local civilian councils, which had failed to make good on their pledges to provide housing for the VVS. To take up at least part of the slack, the VVS committed 80 percent of its capital construction funds in 1991 for family housing. It also established a Main Engineering

---

[120]Interview with Lieutenant General Grigory K. Dubrov, "Who Will Be a Pilot In the CIS's Sky?" *Krasnaia zvezda*, March 24, 1992.

[121]Colonel (later, Major General) A. Tsalko, "We Do Not Value Thrift," *Krasnaia zvezda*, January 24, 1989.

Administration to accelerate the resolution of the housing problem.[122] Yet the following year, some 22,000 VVS families remained without living quarters. Over 3500 of these were families of pilots.

Even for those VVS families lucky enough to be blessed with adequate living accommodations, daily existence is all too often bleak. Shortly before the coup, the VVS's chief political officer noted that around half of all officers' wives possessed special work qualifications, yet lacked any realistic chance of finding gainful employment in the often remote parts of the country where their husbands were stationed.[123] Such deprivation has had a predictable impact on morale. Acknowledging that many officers have remained hardworking and devoted professionals in the face of mounting adversity, the deputy commander of the flight school at Chernigov confessed that "one feels frankly ashamed to reproach people for their deficiencies" when they sit at their work stations on air bases for up to 12–14 hours a day.[124] Much the same sentiment was reflected in a defense ministry poll of 1100 officers in all of Russia's services, including the VPVO and VVS, indicating that many "are losing their social and moral reference points and values, and their confidence in tomorrow is dying away."[125]

At the time of the coup, a Moscow bus driver typically got paid more than a trained Soviet fighter pilot. Since then, the economy has degenerated to a level where line pilots have to work the fields on weekends to help bring in the crop. Many VVS officers have been forced to harvest their own agricultural produce, with base commanders cultivating plots and maintaining subsidiary farms on their airfields. The chief of logistics, Lieutenant General Ivanov, remarked caustically: "We get nothing but extra headaches for this."[126]

---

[122]Interview with Colonel General Shaposhnikov, "So When Will Pilots Get Apartments?" *Krasnaia zvezda*, March 3, 1991.

[123]Interview with Lieutenant General Gennady Benov on the program "I Serve the Soviet Union," Moscow television, August 18, 1991.

[124]Colonel V. Sobolev, "Stop Teaching in a Retrograde Way," *Aviatsiia i kosmonavtika*, No. 3, March 1991, pp. 4–6.

[125]Lieutenant Colonel Nikolai Pechen, "With What Do We Fill the Void?" *Vestnik protivovozdushnoi oborony*, No. 9, September 1993, p. 30.

[126]Interview with Lieutenant General Stanislav Ivanov, *Aviatsiia i kosmonavtika*, No. 9, September 1991, pp. 2–3.

Even at prestigious Kubinka, fighter pilots often spend their spring and summer weekends weeding and hoeing. The former commander of the VVS's Su-27 flight demonstration team, Colonel Vladimir Basov, said that "all of us are forced to tend our kitchen gardens because we don't have any other source of food." He added: "It's a shame our pilots get lower pay than a plumber or a mechanic."[127] General Deinekin himself has commented that cadets at the Barnaul flight school live in such austere conditions that they are forced to use parachutes as blankets during the wintertime.[128]

Again, such problems are not limited to fighter aviation. Colonel General Kalugin spoke candidly of the grim living conditions of Russia's bomber crews. "I visited the flight mess—a very, very poor table. And the families of the fliers? They can barely make ends meet." Kalugin freely acknowledged complaints about social injustice and a lack of legal protection, about nothing to look forward to upon being discharged or retired, and about the persistent "grains of mutual distrust" that had taken root and accumulated over the years.[129]

Sad to say, Russia's pilots have watched their professional pride slowly leach away as a result of pernicious influences like these. Smoking is said to be the rule among them, and drinking to excess has become more and more commonplace. "The whole country drinks, after all, and do they ever!" wrote one disgusted pilot. "Why should aviation be any better?"[130] Only a few officers reportedly take part in regular physical exercise, and many work out only enough to get ready to pass their semiannual evaluation—if it is given. Even these tests are typically a charade because of the widespread prevalence of cheating.

---

[127]Daniel Schneider, "Russian Fighter Pilots Wax Nostalgic for Days of Top Gun Status," *The Christian Science Monitor*, August 5, 1992.

[128]Interview in *Krylia rodiny*, March 1993.

[129]Interview with Colonel General Kalugin, "In the First Strategic . . .," *Aviatsiia i kosmonavtika*, No. 10, October 1991, pp. 2–3.

[130]Lieutenant Colonel V. Vysotskii, "A Stumbling Block, Or Problems of Combat Training," *Aviatsiia i kosmonavtika*, No. 11, November 1991, pp. 4–7. On this point, two VVS flight surgeons reported in 1992 that 80–90 percent of VVS pilots, including fighter pilots, were sedentary, more than half smoked, and roughly 45 percent were overweight. Colonels of Medical Service A. Ivanchikov and V. Chuntul, "Risk Factors and Flight Longevity," *Aviatsiia i kosmonavtika*, No. 1, January 1992, p. 10.

Faltering Service Prestige and Pilot Recruitment

During the banner years of the Soviet Union, appeals to patriotism and the romance of high-performance flight were virtually all it took to entice the best of Soviet youth to seek a VVS career. Today, squalid living conditions and rapidly dwindling opportunities for pilots to fly have become increasing barriers to VVS recruitment.

Consistently low pay for officers and the badly tarnished image of a military career in post-Soviet Russia, set against the precipitous decline in the quality of service life, have resulted in a virtual disappearance of competition for pilot training slots in both the VVS and VPVO. Even before the USSR's collapse, the commandant of the flight school at Kharkov reported that the influx of new cadets had fallen "drastically." He noted that 790 applicants were accepted to Kharkov in 1989, whereas only 312 entered the program in 1990. He added: "There was practically no competition after the medical board's findings. In some cases, we were even forced to reexamine those who received 'twos.'"[131]

During the early 1970s, six to eight applicants typically vied for each available pilot training slot nationwide. Today, the VVS is forced "to accept adolescents who have shown only fair knowledge on the entrance exams. The criterion for their enrollment is just good health, and even that with certain allowances." One colonel complained that "there is essentially no weeding out after psychological testing. There is no one to choose from!"[132] Another pilot cynically joked that flight school acceptance standards had fallen to such a low state that there are now only two criteria: "The applicant must be able to hear thunder and see lightning—and one of these is waiverable!"

---

[131]Interview with Colonel V. Shevstsov, "Pilots Who Never Developed: Can We Halt the Outflow of Cadets from Military Schools?" *Krasnaia zvezda*, March 12, 1991.

[132]Things at the end of 1991 were a little better in VPVO. According to its chief of fighter aviation, then–Lieutenant General Vladimir Andreyev, there was a maximum of 1.5 applicants competing for each position. But this was not much of an improvement over the VVS's situation. General Andreyev said that in 1966, when he entered the VPVO flight school at Armavir, he had already flown 40 hours on light aircraft in the Lugansk DOSAAF aeroclub and that Armavir only accepted applicants with prior DOSAAF training. He also noted that the competition then was seven applicants per slot. Interview by Colonel A. Andryushkov with Lieutenant General Vladimir I. Andreyev, "We Need to Know the Threat By Sight," *Krasnaia zvezda*, November 22, 1991.

Many junior officers have simply quit out of disillusionment. In July 1992, for example, all 48 graduates of the flight school at Barnaul declined to honor their service commitments because of "no prestige and no prospects." Upon being awarded their commissions and aeronautical ratings, they were immediately released into the reserves.[133] In trying to come to honest grips with this disturbing trend, the VVS's chief of education, Major General Yanakov, frankly conceded that "today's youth have begun looking harder and deeper into life's questions. They can no longer be won over simply by slogans and appeals. Firm assurances of a dignified social status of officership are now required."[134]

---

[133]"Military Pilots Are Reluctant to Serve," *Izvestiia*, July 4, 1992.

[134]Major General Ya. Yanakov, "From Retrograde to Modern," *Aviatsiia i kosmonavtika*, No. 1, January 1992, pp. 2–3.

Chapter Three

EVOLVING ORGANIZATION, DOCTRINE, AND FORCES

Under General Deinekin's leadership, the VVS has stepped out aggressively to cope with its many problems and challenges just described. It has managed an unprecedented drawdown of forces in the space of four years. This has included completion of the return of all forward-deployed Soviet Air Force assets in the Baltic states and former Warsaw Pact countries of Eastern Europe. The drawdown has also entailed a reduction in deployed combat aircraft well below the Conventional Forces in Europe (CFE) Treaty ceilings, as well as a summary retirement of many obsolescent and obsolete aircraft.

At the same time, the VVS has moved to consolidate its organization and functions, to increase the quality of its equipment, and to attend to the needs of its people. It has sought a new image to help restore the attractiveness of VVS service, including adoption of a new uniform, with air force blue replacing the old army green. It is searching for new operational concepts appropriate to Russia's still-undefined security challenges in the post–cold war world. And it is pursuing a measured force modernization effort in the face of severe fiscal constraints.

This chapter addresses the VVS's organizational, conceptual, and structural adjustments to the post-Soviet era. Among other things, it seeks to cast light on how General Deinekin's air force figures in broader Russian military reform plans, on changes to date in VVS composition and orientation, and on trends in VVS force structure and force modernization. Although the main focus is on the VVS, the separate Air Defense Forces (VPVO) and naval aviation are touched upon briefly.

On what basis can one venture informed statements about such questions? Like the preceding chapter, the analysis that follows here is based on the unprecedentedly rich information provided by the Russian military press since the advent of glasnost, as well as on numerous published interviews with senior VVS and defense establishment leaders. These individuals, who have little incentive to lie to us now that the cold war is over, have been uncharacteristically forthcoming about Russia's force development ambitions and the various constraints upon them. Increasingly, as both the VVS and Russia's aviation industry have been driven to compete for scarce funding in the bureaucratic-politics mode familiar to Western defense establishments, their leaders have become more open about their plans and problems out of natural vested self-interest.

To characterize the extent of reductions in the short period of a few years, the present chapter also includes several time-series depictions of the devolution of Russia's air power from the final days of the Soviet juggernaut in the late 1980s to the badly truncated forces inherited by Moscow in the wake of the USSR's demise. The data supporting these depictions are derived from figures contained in the 1989–1994 editions of the annual *Military Balance* report published in London by the International Institute for Strategic Studies, as updated and refined wherever possible by more authoritative information provided in CFE Treaty documentation and in public disclosures by Russian defense officials. The resultant charts make no pretense of offering precise numerical and descriptive comparisons between where Russia's air power strength was in the late 1980s and where it stands today. However, the data on which they are built are more than good enough to portray the broad trend in Russian air power since 1991 as one of steep and continuing decline.

Finally, the discussion draws liberally on a remarkable document compiled in 1994 by the VVS's Central Research Institute and edited for publication in a Western symposium volume by the respected Russian civilian defense expert, Dr. Aleksei Arbatov, formerly of the Institute of World Economy and International Relations and now head of the Subcommittee on International Security and Arms

Control of the Russian State Duma.[1] That document, referred to hereinafter as "the VVS analysis," is unprecedented in Russian practice. In both breadth and depth, it comes as close as anything to date to being the Russian equivalent of a U.S. military posture statement. It is astonishingly frank in describing the VVS's roles and missions, current status, and force development plans to the year 2015, as well as the many problems that threaten to obstruct the implementation of those plans. Even in the recent past, such information would have been treated by the Russian defense bureaucracy as highly sensitive. Its public release in this form offers impressive proof of Russia's fitful trend toward greater openness in the military sphere. Beyond that, the document is an exemplary testament to the sort of cooperative work that can be done jointly by Russian and Western defense professionals.

POST-SOVIET RETRENCHMENT AND REFORM PLANS

In the immediate aftermath of the collapse of the Soviet state, Russia strove to maintain an integrated military posture throughout the newly created Commonwealth of Independent States (CIS). Once that goal proved evanescent, Marshal Shaposhnikov, in his new role as commander in chief of the CIS Joint Armed Forces, continued to plead without success for the preservation at least of joint CIS air and air defense forces, since those had been designed and fielded with the former USSR's strategic defense needs in mind rather than with regard for the new borders that had appeared as a result of the union's collapse.[2] Shaposhnikov also pled, likewise to no avail, for an alliance of five or six CIS core states modeled on NATO.[3]

Once the CIS heads of state failed to agree on defense integration at their crucial meeting in Minsk on February 14, 1992, it was only a matter of time before an independent Russian military, including a

---

[1]"Russian Air Strategy and Combat Aircraft Production: A Russian Air Force View," in Randall Forsberg, ed., *The Arms Production Dilemma: Contraction and Restraint in the World Combat Aircraft Industry*, Cambridge, Massachusetts, MIT Press, 1994, pp. 17–60.

[2]ITAR-TASS, October 12, 1992.

[3]Interview by Marina Churnukha, Interfax, November 20, 1992.

Russian Air Force, would be created from the detritus of the USSR. As the outgoing chief of the Soviet General Staff, General Vladimir Lobov, observed, "the process of disintegration of a once-unified organism is becoming more and more irreversible," occasioning "an insistent necessity . . . of forming and putting into action a Ministry of Defense and a General (or Main) Staff for a Russian military in the near future."[4]

As early as January, a Yeltsin decree anticipating creation of a Russian defense ministry was drafted.[5] Soon thereafter, efforts to proceed with developing an independent Russian defense establishment gained momentum.[6] Yeltsin's newly appointed defense minister, Army General Grachev, conceded that it could take as long as ten years for Russians "to be able to speak with full confidence about the establishment of Russian Federation armed forces in a new guise," more or less the same length of time he felt would be required for Russia's new statehood to reach maturity.[7] Four major hurdles cited by Grachev included establishing a new and smaller force posture "consistent with the times," making a smooth transition to new hardware based on modern technology, fundamentally reforming existing concepts of training and force employment, and creating a new image for the Russian serviceman.

The Formation of a Russian Military

The decree setting up the Russian armed forces was signed by President Yeltsin on May 7, 1992. In August, the service chiefs and other senior officials of the Russian defense ministry and armed forces were announced. Colonel General Vladimir Semenov was ap-

---

[4]General Vladimir N. Lobov, "The Sacred Right of Russia," *Sovetskii patriot,* No. 8, February 1992, p. 4.

[5]Pavel Felgengauer, "The Army Is Under Fire from Politicians," *Nezavisimaia gazeta,* January 11, 1992.

[6]A more comprehensive review of this background is presented in Roy Allison, *Military Forces in the Soviet Successor States,* Adelphi Paper No. 280, London, International Institute for Strategic Studies, 1993, especially pp. 3–35. On broader strategy and foreign policy developments, see also Renee de Nevers, *Russia's Strategic Renovation,* Adelphi Paper No. 189, London, International Institute for Strategic Studies, 1994.

[7]Petrovskii, *Rossiiskiye vesti,* January 4, 1993.

pointed commander in chief of the Ground Forces; Colonel General Igor Sergeyev, commander in chief of the Strategic Missile Forces; Admiral Feliks Gromov, commander in chief of the Navy; and Colonel General Viktor Prudnikov, commander in chief of VPVO.[8] General Deinekin was named commander in chief of the new Russian VVS less than two months thereafter.[9]

As a first item of business, Grachev reported that the decline in readiness of Russia's forces had been arrested, that command and control had been restored, and that an initial inventory of the equipment the new Russian military inherited had been conducted. His immediate plan was to start reducing the size of the Russian military by a quarter, to three million troops, in compliance with the negotiated terms of the CFE Treaty.[10]

In June 1992, Grachev announced a military reform program that would proceed in three phases. The first phase, to be completed in 1993, included establishment of a Russian defense ministry; laying a legal foundation for the creation of Russian Federation armed forces; initial steps at defining the organization, structure, and force levels of the Russian armed forces; withdrawal of troops under Russian jurisdiction from the former Soviet republics; and the development of a command and control structure for the Russian armed forces.

The second phase, to be completed in 1994, would close out the withdrawal of Russian troops from the former republics; further reduce and restructure Russia's remaining forces; establish a system of social safeguards for servicemen and their families; and shift to a mixed conscript and voluntary base of recruitment, with provision for alternative national service. Grachev indicated that throughout this second phase of reform, the existing arrangement of five military services (Strategic Missile Forces, Ground Forces, VPVO, Air Force, and Navy) would be retained.

---

[8]"Russian Federation Defense Ministry Press Center Reports New Appointments," *Krasnaia zvezda*, August 22, 1992.

[9]ITAR-TASS report, October 6, 1992.

[10]Michael Parks, "Soviet Army Will Be Cut, Says New Defense Minister," *Philadelphia Inquirer*, September 21, 1991.

For the third phase, beginning in 1995, Grachev allowed for the possibility of reorganizing and combining various services. He stressed, however, that any such changes would be carried out in an evolutionary manner, since breaking up existing structures over the next two to three years, while Russia remained in the grips of a severe economic crisis, could trigger a breakdown of order and discipline, a lowering of readiness, and perhaps even "the utter disintegration of the military."[11]

The third stage of Grachev's planned reforms, to continue until the end of 1999, will see a reduction in troops by 1.3 million, bringing the military down to 1.5 million overall, with defense expenditure ideally stabilized at around 5–6 percent of gross national product.[12] Also during this third stage, open season will be declared on the existing five-service arrangement, including much realignment and consolidation of assets and perhaps some services disappearing altogether, with VPVO's fighters going to the VVS and the rest of its assets to a new Strategic Deterrent Force. Ground force organization will then shift from an army/division focus to a corps/brigade focus, with ensuing implications for air support needs.[13]

Grachev openly chafed at the understandable difficulty of conducting rational defense planning "in the absence of an elaborated, officially adopted Russian military doctrine."[14] No doubt in part to help alleviate that concern, such a doctrine was finally promulgated in November 1993 by the Ministry of Defense, after much anticipation

---

[11]Lieutenant Colonel A. Dokuchayev, "The Russian Armed Forces: Phases of Formation," *Krasnaia zvezda,* June 23, 1992.

[12]Interview with Grachev, "A Strong Army Heightens the Self-Respect of the People," *Armiia,* No. 11–12, 1992, pp. 3–8. Grachev also reaffirmed Russia's commitment to a nuclear no-first-use pledge, a commitment later retracted in the new Russian military doctrine. Whether the promised 1.5 million troop ceiling will hold remains to be seen. It has come under fire from many of diverse persuasions within the military, who have argued that a troop level of 1 percent of the population is insufficient for a country of Russia's size and stature.

[13]Colonel O. Falichev, "Building the Russian Military Is a Task for All the People," *Krasnaia zvezda,* May 26, 1992.

[14]See "Army General Pavel Grachev: Clear Calculation and Common Sense Are Needed in Forming the Russian Military," *Krasnaia zvezda,* July 21, 1992.

both in Moscow and in the West.[15] It contained numerous warmed-over elements of the old Soviet doctrine for conventional warfare, with appropriate amendments at the margins to account for Russia's new regional security preoccupations. Quite unlike the former Soviet doctrine, it was also prefaced by a formal declaration that Russia has no enemies.

Marshal Shaposhnikov was among the first to criticize the issuance of such a doctrine when some important prior questions remained unanswered. "We still do not know," he complained, "what we are, where we are going, and what our ultimate goals are. . . . Our blueprint for national security should follow from a blueprint for the development of the Russian state. . . . We have to say: These are our interests, these are the possible dangers and threats to our interests, and from this you get a blueprint, a doctrine. . . ."[16] He had a valid point.

Toward a Mobile Force Concept

While such questions continue to be sorted out in the ongoing Russian defense debate, Grachev's reform plan envisages a new military consisting primarily of a rapid reaction force made up of airborne units, lightly armored and air-transportable motorized rifle formations, Mi-26 helicopters, and military transport aircraft, plus marines and logistic support units.[17] Amplifying on this planned mobile force concept, Grachev said that the biggest structural changes would affect the ground forces, to include the creation in each military district of several divisions at full readiness, with the rest being considered reserve divisions.[18]

---

[15]For a balanced analysis, see Charles J. Dick, "The Military Doctrine of the Russian Federation," *Journal of Slavic Military Studies*, No. 3, September 1994, pp. 481–506.

[16]Interview on Radio Rossiia, November 21, 1993.

[17]Interview by Sergei Ovsiyenko with General Pavel Grachev, "The Army: Overcoming Arrhythmia," *Rossiiskiye vesti*, May 29, 1992.

[18]Moscow First Program television interview, May 31, 1992. For a more detailed treatment of post-Soviet military reform plans and their prospects, see Richard F. Staar, *The New Military in Russia: Ten Myths That Shape the Image*, Annapolis, Naval Institute Press, 1996.

Shortly before his selection to become First Deputy Minister of Defense, Andrei Kokoshin observed that the long-dominant armored component of the Soviet armed forces had become "an anachronism, a dinosaur from World War II." He suggested that the new Russia needed to rebuild its military with primary emphasis on the high-technology services (VVS and VPVO, Strategic Missile Forces, and the Russian Navy) and with "significantly reduced and restructured ground forces" configured for rapid deployment to any area where an outside threat to Russia's security might arise. Kokoshin added that Russia should "not rely too much on a nuclear shield," but rather should concentrate its attention and resources on building up small but efficient general-purpose forces.[19] Unfortunately, legitimate concern has since arisen that Russia will eventually be driven to rely all too heavily on a nuclear shield to compensate for its ever-worsening inadequacies in conventional strength.

One of the first hints of official thinking about the likely composition and character of Russia's mobile forces to deal with peripheral threats came in a defense ministry announcement in November 1992 that such forces would begin as an interbranch, or joint (*mezhvidovoi*), combat formation and might later become an autonomous component of the armed forces. It would center on the inland Volga and Urals Military Districts. With the collapse of the Soviet war machine and the persistence of lengthy Russian borders, it was no longer possible to guard these borders with permanent garrisons, thus necessitating the development of mobile forces.

The core of this new peacekeeping force, envisaged as operating under CIS and United Nations auspices, would be made up of Russian airborne troops. During its formative stages, it was envisaged by Grachev as including two airborne divisions and three airborne brigades, backstopped by several army helicopter regiments, three

---

[19]Interview by Yelena Agapova, "Before You Form an Army, You Should Know What It Is For, Expert Andrei Kokoshin Believes," *Krasnaia zvezda*, March 17, 1992. No first use of nuclear weapons was always a Soviet propaganda ploy. One of the many skeletons let out of the former Soviet strategic closet was a disclosure by Germany's defense minister, Gerhard Stoltenberg, citing former East German archives, of a Warsaw Pact contingency plan to carry out a preemptive strike with 840 nuclear warheads to smash NATO opposition and break through NATO's defenses in case of war in Central Europe. See Mikhail Shchipanov, "Will There be Secrets from Our Ally?" *Kuranty*, February 7, 1992.

marine battalions, and some VPVO and communications units, along with subordinated VVS fighter and ground attack squadrons—all of which would begin to take shape in concrete form sometime in 1995. Ultimately, the VVS component of Russia's rapid deployment force would comprise five or six fighter regiments, five bomber regiments, two ground attack regiments, and four airlift divisions. That was, of course, before Russia's invasion of Chechnya, with its ensuing costs and complications, put any serious thoughts of further reform on indefinite hold.

The restructured Ural and Volga Military Districts were elevated in importance from rear to second-echelon districts as a direct result of the USSR's disintegration. They were chosen to provide a real-world setting in which the High Command might validate its sought-after shift from the large battle formations of classic Soviet military practice to more highly mobile, rapid-reaction forces. The first echelon of these newly configured forces was scheduled to be based in the North Caucasus Military District, which was said to confront the main threat of ethnic rivalries to the south. As it turned out, that was a prescient call, since the Russian invasion of Chechnya two years later drew heavily on military forces, particularly combat aircraft assets, that had recently been relocated to the North Caucasus Military District.

NEW LOOKS FOR THE RUSSIAN AIR FORCE

General Deinekin's Russian Air Force of today is but a faint shadow of the former Soviet Air Force he inherited from Marshal Shaposhnikov four months before the collapse of the union in December 1991. As noted earlier, most of the latest-generation combat aircraft of the Soviet Air Force, as well as its most well-developed airfields, had been positioned beyond the westernmost borders of the Russian Federation. Those assets were lost to the newly independent states once the former union ceased to exist. Russia retained a large number of jet trainers and earlier-generation combat aircraft, along with their associated bases. However, the VVS lost 37 percent of the former Soviet Union's MiG-29s, 23 percent of its Su-27s, 43 percent of its most modern Il-76 jet transports, and the overwhelming majority of its Tu-160 and Tu-95 strategic bombers.

In all, barely 60 percent of the aircraft and 50 percent of the air bases of the former Soviet Air Force were left on Russian soil.[20]

About 40 percent of the VVS's depot-level maintenance and repair facilities were also lost to the newly independent states. Its air-to-ground missiles had been inspected and maintained solely in Estonia. More than half its Su-24s had undergone phase maintenance in Lithuania. Its Su-25s had been serviced only at depots in Lithuania and Georgia, and its Tu-95s had been overhauled exclusively in Ukraine. Because of the severe shortage of assemblies and spare parts in Russia, the VVS has no near-term solution to the problem these losses have created. Among its most modern bases, the VVS lost 44 in Eastern Europe and 94 to the former Soviet republics, leaving Russia with some 90 major airfields altogether, only half of which are top quality. Even the latter need considerable enhancement with shelters and better maintenance and support facilities.[21]

Force Reductions and Consolidation

With no funding to support the procurement of new aircraft, General Deinekin's near-term options for replenishing these losses were limited to withdrawing what little current-generation equipment the VVS had deployed in the Baltic states, Poland, East Germany, and Transcaucasia. That withdrawal did not begin with the collapse of the USSR. On the contrary, the Soviet Air Force had initiated a planned pullback of forward-based regiments, divisions, and air armies from Czechoslovakia and Hungary as early as 1990. In 1991, it commenced a withdrawal of aircraft and units from Poland, the Baltic states, and Transcaucasia, and in 1992 from East Germany.

---

[20]Interview by Piotr Butowski with Colonel General Anatoly Malyukov, "Flying in the Face of Adversity," *Jane's Defense Weekly*, April 17, 1993, p. 15.

[21]The VVS was by no means the only service so afflicted. In a speech to the Congress of People's Deputies in December 1992, Grachev starkly characterized the dregs that Russia found itself left with following the disintegration of the former Soviet armed forces: "Ruins and debris, basically. Communications, command and control, intelligence, missile attack warning, air defense, and logistical support systems were wrecked. We inherited forces from the second strategic echelons. The most combat-capable units, equipped with the latest armaments, were left outside the Russian Federation. Thousands upon thousands of Russian servicemen and their families ended up outside." *Krasnaia zvezda*, December 8, 1992.

Altogether, some 300 operating units, more than 30,000 VVS personnel, and 700 combat aircraft were brought home to Russian territory during that period, with more than half those units and over 500 aircraft redeployed to the air forces of the Moscow Military District alone.

In 1992, 36 air regiments were withdrawn to Russia from Eastern Europe and the former Soviet republics.[22] In 1993, the VVS further withdrew its 40 MiG-23s and ground support personnel deployed at the Burevestnik airfield on Iturup, one of several islands in the Kuril chain that Japan claims as its territory. That move left Russia with no remaining air bases in the disputed region.[23] By late 1994, 40 regiments had been withdrawn from former forward operating locations altogether.[24]

Russia's aviation manpower has also undergone a significant decline in strength from its former Soviet level. From a total of a little over a million men in the VVS, VPVO, and naval aviation in 1989, the combined number for the three air arms today is down to barely more than 500,000 (see Figure 1).[25] Furthermore, as a result of the failure of conscription, the three services have become abnormally top-heavy with officers. In 1989, 69 percent of the VVS and 60 percent of VPVO were conscripts. Today, by contrast, officers and career noncommissioned officers in both services outnumber conscripts two to one.

---

[22]Interfax, November 28, 1992.

[23]"News Breaks," *Aviation Week and Space Technology,* August 2, 1993. This could have come only as good news to the VVS pilots stationed there. The regiment at Iturup was the only one from which Russian pilots flew single-engine fighters over water. As a press comment somberly noted, "the first failure of a MiG-23's engine automatically becomes the last failure in the pilot's life. A ship or submarine will arrive at the accident site three days later at best." Burevestnik was described as a bare-base facility, with no hangars and with aircraft "rusting year-round under the open sky." I. Kots, "Islands in Shoulderboards: Whom Is the Military Deterring in the Southern Kurils?" *Komsomolskaia pravda,* July 28, 1992.

[24]Interview with Colonel General Deinekin, "Russia Will Not Remain Without Wings," *Krasnaia zvezda,* November 5, 1994.

[25]The figure for VPVO includes personnel assigned to surface-to-air missile and radar units in addition to fighter aviation.

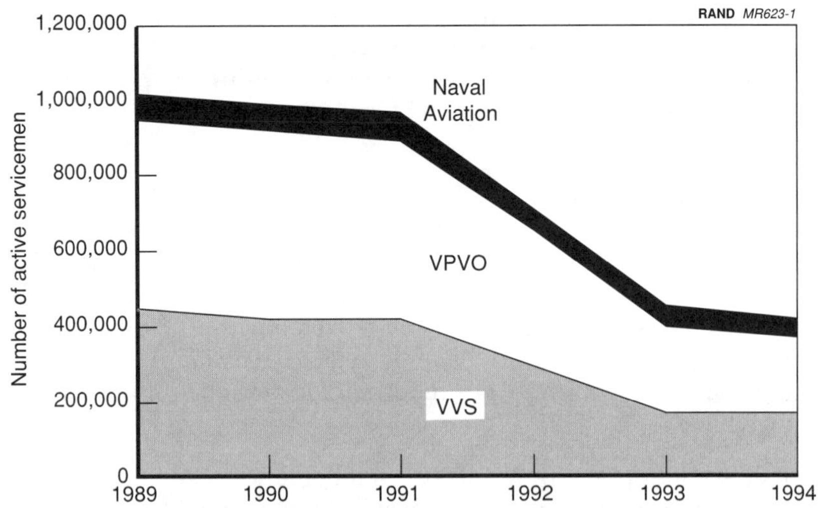

Figure 1—Aviation Manpower, 1989–1994

Employment Concepts and Air Doctrine

In traditional Soviet military doctrine, the army-dominated General Staff subordinated air power to a secondary role as a supporting element in a combined-arms approach to war fought and won mainly by massive infantry and armored forces. Almost overnight, the operational focus of the Russian armed forces shifted from a stress on theater offensive warfare to regional power projection, which naturally played to the greatest strengths of the VVS. Increasingly since the success of coalition air power against Iraq in Operation Desert Storm, Russian military experts, and not only airmen, have come to recognize and accept the ascendant capabilities of aviation.[26]

---

[26]This view has not been universal. As one might expect, the chief of the ground forces headquarters staff conceded that although aircraft in Desert Storm "showed themselves to be a maneuverable, effective weapon" and played the leading role in that particular war, this in no way "belittles the significance of ground forces' military actions in the attainment of the ultimate goal." Indeed, he said, the war "confirmed the philosophical tenet of their determining role." Interview with Lieutenant General Yury D. Bukreyev, "I Favor the Ground Forces," *Krasnaia zvezda*, November 28, 1991.

Tacitly acknowledging the example of Desert Storm, the VVS analysis cited at the beginning of this chapter holds that "the success of ground operations increasingly depends on air force missions, from achieving air supremacy until the moment the enemy surrenders." It further echoes the well-known refrain among Western airmen that the essence of air power is flexibility, calling such flexibility "crucial for the defense of Russian territory, with its vast expanse, decreasing inventory of combat aircraft, urgent need to allocate economic resources to support and maintenance facilities, and uncertainty about the potential origins and combinations of future threats."

The VVS analysis concedes that the likelihood of major war has diminished with the ending of the cold war and that both Russian and American planners now believe the chief near-term danger to entail wars of low and medium intensity. Revealingly, however, it adds that the chance of high-intensity conflict cannot be ruled out altogether, since "low-intensity conflicts are not sufficiently demanding to define the size or technology of the Russian Air Force." This reservation testifies unabashedly to the persistent determination of the VVS to continue pressing the state of the art in aviation technology, even though Russia confronts no near-term threats that would remotely warrant new systems going beyond the capability of the weapons Russia already possesses, let alone the expenditure of scarce funds that would be needed to pay for such systems. In a discerning comment on this aspect of the analysis, Aleksei Arbatov notes that its assessment of the strategic landscape and the more implausible contingencies it cites as a basis for Russian force planning reflect the natural bureaucratic tendency of any military organization "to retain as much as possible of its traditional strategic roles and operational missions, giving only lip service to the new post-cold war security realities."[27]

The VVS analysis sketches out five scenarios that, it says, constitute the core of the planning assumptions assigned to the VVS by the defense ministry from now through the early years of the 21st century:

The Northwestern and Western Strategic Salient. In an apparent reach for the one "high intensity" scenario needed to justify continu-

---

[27]See his "Editor's Comment: Military Reform and the Air Force," in Forsberg, ed., *The Arms Production Dilemma*, pp. 51–60.

ing with a robust R&D and force development program, the analysis cites the possibility that "NATO might try to employ force to settle Russian internal conflicts, to deny Russia its legitimate interests, or even to seize parts of its territory to undercut strategic positions or for post-conflict bargaining." This fanciful scenario postulates that NATO would begin any such offensive with intense air and naval bombardment aimed at seizing the Kaliningrad region and then would press to Russia's western frontier through Belarus and Ukraine, employing both air attacks and deep ground-force penetration into the Leningrad and Moscow Military Districts. This possibility, according to the analysis, requires a VVS capability to repel enemy air operations, prevent amphibious landings, and conduct offensive and defensive counterair operations over enemy territory.

The Southwestern Strategic Salient. Somewhat more plausibly, the analysis visualizes subtle efforts by Turkey and Iran to weaken Russia's position in the Transcaucasus and to win over the largely Muslim populations of Azerbaijan, the North Caucasus, and several other conflicted areas in the region. This scenario suggests that an escalation of fighting between Armenia and Azerbaijan, or the complete disintegration of Georgia, might lead to large-scale military intervention by Russia, Turkey, and Iran. Such a possibility could harbor all sorts of escalatory potential, including a Turkish attack supported indirectly by NATO.[28] The latter threat, suggested the analysis, would necessitate prompt counteroffensive operations by the North Caucasus air group and, later, by Russian combined-arms formations. (Interestingly, and perhaps revealingly, the analysis anticipated *no* military and air operations such as those subsequently carried out by Russian forces against Chechnya.)

The Southern Strategic Salient. Similarly, the analysis conjures up the prospect that Turkey, Iran, Afghanistan, and Pakistan could seek to play on Muslim sympathies in Central Asia and Kazakhstan and seize strategically important areas and assets in the region. This

---

[28]The seeds of such a scenario certainly exist among those secessionist leaders of potential breakaway Muslim republics in southern Russia who would seek aid and comfort from their Muslim brethren in Turkey. Dzhokar Dudayev, the former Soviet Air Force major general who led the rebellion in Chechnya, appealed to a Turkish reporter in late December 1994 for Ankara to supply him with combat aircraft so he could use them to bomb Moscow. See Sonni Efron, "Yeltsin Defends Continuation of Chechnya War," *Los Angeles Times*, December 28, 1994.

would require an opposing coalition of Tadzhikistan, Uzbekistan, Kazakhstan, and Russia, spearheaded by VVS aircraft from the Caucasus and Volga-Ural military districts, to provide air cover, tactical reconnaissance, and interdiction of enemy lines of communication.

The Far Eastern Strategic Salient. This scenario reflects continued Japanese claims on the Southern Kuril islands and a stated concern that Russia's nearly complete demilitarization of the Southern Kurils might embolden Japan to solve its "Northern Territories" problem by force. It compounds this extreme implausibility by envisaging Japan as conducting such an operation with active U.S. complicity. It then erects guaranteed preconditions for a self-fulfilling prophecy by suggesting that the VVS would be obliged to engage American aircraft carriers in support of any Russian counteroffensive.

The East Siberian Strategic Salient. In what comes perhaps closest to being a valid worst-case situation that Russia might have plausible grounds to worry about, the analysis contemplates a possible Russo-Chinese war triggered by unrenounced Chinese territorial claims against portions of Kazakhstan, Kirgizstan, Tadzhikistan, and Russia. This last scenario acknowledges that Russia's long border, insufficient ground forces, lack of strategic depth, and exposed infrastructure and lines of communication in the region would put a special premium on early and intense VVS counterattacks against high-value Chinese military, command and control, and industrial targets.[29]

In considering the force requirements for these scenarios, the analysis points out that the VVS would perform numerous functions generic to all circumstances, yet that each scenario presents unique demands as well, considering the wide variations in their respective climate and topography, basing and maintenance infrastructure, prospects for prompt reinforcement and resupply, and likely opposing forces. General Deinekin almost surely had such variations in mind when he said that "we are working out definite views on waging armed conflict depending on the theater of military operations and the forces situated there," adding as an example that if the hypothet-

[29]For a thoughtful independent Western view on the plausibility of this scenario, see S. Enders Wimbush,"When China Absorbs the Russian Far East," *Wall Street Journal*, April 25, 1996.

ical enemy were in the Far East, then "one must take into account the concept of enormous space." Different looks would be required for operations in mountainous areas.[30]

The analysis was candid in conceding VVS limitations. In the Far Eastern scenario, it acknowledged the small number of airfields in the region as a major constraint on dispersal, redeployment, and reinforcement. It further conceded that any Russian air operations against Japan would be of "quite limited effectiveness," considering the sophistication of Japanese and American air defenses and the VVS's shortage of combat aircraft with sufficient range.

As for the "high-intensity" case of Eastern Europe, the analysis frankly admitted that the VVS "would be at a disadvantage." NATO's air forces, it noted, would be able to deploy quickly to forward bases in Eastern Europe, and NATO "generally has longer-range and better air-to-surface missile attack capabilities." To accommodate any such contingency, the VVS would mainly use fighters in conjunction with VPVO to protect Russia, since the VVS's ground attack aircraft suffer from short ranges and limited in-flight refueling capability. The analysis further acknowledged that any use of Russian medium- and long-range bombers would result in "unacceptable" attrition, since NATO's air defenses are dense and Russia's fighter escorts lack sufficient endurance to accompany the bombers to their targets.

The analysis concluded that the VVS's current force structure, aircraft and weapons mix; industrial support; deployment pattern; basing and maintenance infrastructure; command, control, communications, and intelligence (C3I); and redeployment capability are "woefully inadequate to respond to the regional contingencies of the new [Russian] military doctrine." What it lacks to match this commendable candor, unfortunately, is an awareness that its insistence on clinging to outmoded threat constructs in the interest of justifying extravagant force modernization goals will only work to postpone both the VVS's recovery to good health in an age of continuing re-

---

[30]Colonel General Deinekin, "Former Soviet Strategic Bombers May Turn Into Scrap Metal," reprinted in R. W. Dellow, *Organization and Equipment: Priorities for the Russian Air Force*, Conflict Studies Research Center, Royal Military Academy Sandhurst, Camberley, England, June 1993, p. 8.

source constraints and the broader gains to be had from a cooperative security relationship between Russia and the West.

New Commands and Organizational Arrangements

General Deinekin has touted the VVS as the chief provider of mobility for Russia's armed forces. With that in mind, he has declared that the main goal of VVS restructuring through the year 2000 will be to create, from existing formations, "a separate, highly mobile branch of the armed forces" featuring an appropriate mix of personnel, platforms, and weapons able to perform the full spectrum of combat missions either jointly or independently.[31]

To infer from such planning that Russia's air power is being "fragmented" and thus done out of any chance "to reach its full potential"[32] is to mistake its motives. On the contrary, General Deinekin has moved to consolidate the operational components of the VVS into four major commands: Long-Range Aviation (LRA), Military Transport Aviation (*Voenno-transportnaia aviatskiia*, or VTA), a new Frontal Aviation Commmand (*Komandovanie frontovoi aviatsii*, or KFA), and a new Reserve and Training Command (see Figure 2). The first two are familiar holdovers from the Soviet era. The latter two represent an attempt to gain greater coherence and efficiencies in VVS organization.

According to General Deinekin, this step was taken in part to help the VVS reduce the size of its educational and training establishment and its management superstructure.[33] More important, however, is the firmer grip the new Frontal Aviation Command gives the VVS over its fighter and ground attack assets. Previously, Russia's tactical air power was subordinated to the regional military districts under the immediate operational control of the ground forces. The new com-

---

[31]Colonel General Petr S. Deinekin, "Always Ready for Action," *Krasniy voin*, August 7, 1993, p. 3.

[32]Major Brian Collins, "Russia Fragments Its Air Power," *Air Force Magazine*, February 1994, pp. 62–65.

[33]Interview with Colonel General Deinekin, "The New Russian Air Force," *NATO's Sixteen Nations*, No. 3, 1993, p. 39.

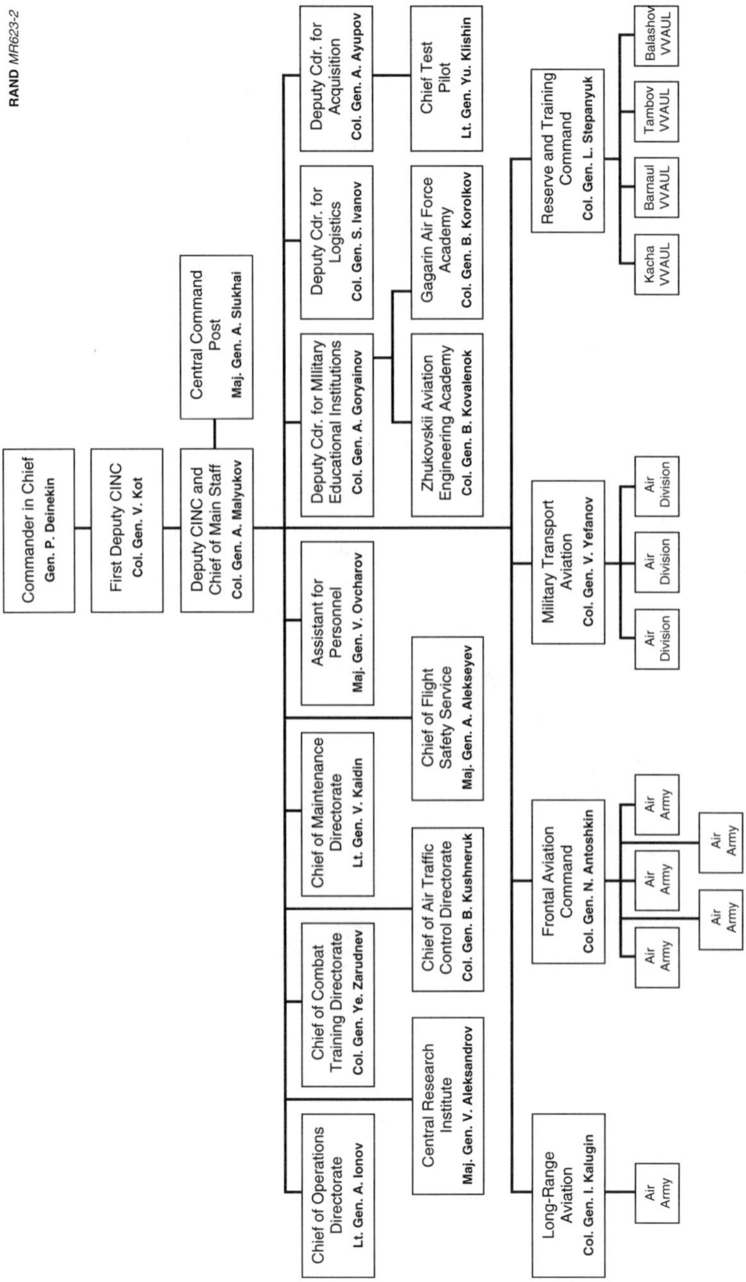

Figure 2—VVS Organization

mand set up to rectify this situation was inaugurated in February 1994 under Colonel General Nikolai Antoshkin, who for the preceding five years had been the air commander for the Moscow Military District.

General Deinekin has also moved to eliminate four former district directorates in an attempt to cut back on the VVS's administrative overhead. In its new incarnation, the scaled-down VVS will have only 170 generals instead of 300, with numerous former officer billets remanded to civilians. General Deinekin said that this reorganization would be carried out in several stages en route to its completion at the end of 1995, with a view toward making the VVS "the most important constituent of Russia's armed forces, as is happening in all the world's developed countries."[34] Whether the promised reorganization was completed on schedule has not been indicated in subsequent VVS pronouncements. It is apparent, however, from Russia's subsequent combat experience in Chechnya that the VVS has a ways to go yet before it can lay claim to being "the most important constituent" of the Russian military (see Chapter Eight for more on this subject).

According to General Deinekin, the VVS and VPVO, while still independent services, have made a concerted effort to retain in the post-Soviet era the interactivity that existed between them throughout the Soviet period since VPVO's establishment as a separate service in 1954. General Deinekin added that gaps in the coverage of Russia's surface-to-air missile defenses created by the breakup of the USSR had increased the burden on VVS fighter aviation, which will use its capabilities "to the maximum" in support of the home defense mission. On the bureaucratically sensitive issue of VPVO's future as a separate service, General Deinekin averred that the imperatives of reform "could require, at a certain stage, combining the forces and means of the VVS and VPVO into a single service branch." He added with diplomatic grace, however, that any such move would be a "complex and painful process" involving diverse pros and cons.

On the positive side, Deinekin rightly observed, such a merger would permit more centralized command and control of air defense opera-

---

[34]Interview with Colonel General Deinekin, "Russia Will Not Remain Without Wings," *Krasnaia zvezda*, November 5, 1994.

tions, along with greater efficiencies and resultant economies from consolidating duplicative airfields and maintenance and logistics systems. The problems associated with any such merger, he suggested, would stem mainly from the fact that the two services have traditionally been separate, necessarily implying difficult changes in institutional practices and habit patterns. He added that considerable financial problems would be created by any such consolidation during the transition period. Accordingly, Deinekin suggested that neither the Yeltsin government nor the defense ministry was yet ready to proceed with combining the two services, if only because of the additional burden on state funds it would impose in the short run. More recently, Grachev indicated that VPVO will be retained, at least for the near term, as a separate service.[35] Most Russian defense officials, however, including the senior VPVO leadership, would probably agree that VPVO's days as a separate service are numbered in the longer run.

New Basing Approaches

An acute problem for the VVS involves rationalizing its inherited deployment and basing infrastructure. Its current deployment pattern is wholly a reflection of what Russia was left with after the USSR fell apart, bearing no relationship to subsequent contingency planning, intended VVS reforms, or military doctrine. The VVS analysis notes that around 70 percent of the VVS's aircraft are based in the European part of Russia, with 15 percent located in the northwestern direction, 25 percent in the western direction, and 30 percent in the southwestern direction. Of the remaining 30 percent east of the Urals, 20 percent face the Far East and only 10 percent face China. The analysis properly calls this inherited deployment pattern "distorted" and points out that the VVS needs fewer aircraft in the Northwestern and Far East salients and more in the North Caucasus Military District opposite Transcaucasia, Turkey, and Iran.

The analysis further states that two-thirds of all VVS aircraft in European Russia are based only 200–300 km from Russia's western borders, and that bombers and transports are concentrated on an

---

[35]Aleksandr Ivanov, "VPVO Forces Retained as a Branch of the Armed Forces," *Krasnaia zvezda*, April 12, 1994.

"unacceptably small" number of bases. The VVS faces a severe ramp space problem owing to the massive withdrawal of former Soviet aircraft from the forward area over the past three years. According to the VVS analysis, only half of Frontal Aviation's fighters are now revetted or protected in hardened shelters.[36]

Russian analysts have noted with interest the USAF's development of composite wings and its application of the composite-wing concept during the 1991 Persian Gulf War.[37] It remains to be seen whether the VVS will adopt an analogous approach, although a precursor of sorts might be the 234th Proskurovskii Guards Fighter Regiment at Kubinka near Moscow. This wing-equivalent consists of MiG-29s, Su-27s, Su-24s, and Su-25s. What makes it unlikely as a composite-wing prototype is that Kubinka has long been a showcase air base for visiting foreign dignitaries and the home of the VVS's flight demonstration teams. There are no similar regiments at other VVS bases.

At the same time, the VVS analysis has advocated a new approach which it calls "aircraft basing regions," aimed at providing a fully equipped operations and maintenance infrastructure for all aircraft permanently based in the region, as well as for any additional types that may be deployed to a given direction during mobility exercises or actual crises. These bases are to be established at existing Frontal Aviation, LRA, and VTA major airfields located more than 300 km inside Russia's new borders. According to the analysis, these bases will be configured to maintain and support all aircraft types earmarked for operational roles in their assigned regions. They will be designed and equipped to accommodate five or six permanently based fighter squadrons, and will serve as hubs for outlying airfields attached to them. The analysis concedes that the biggest obstacle threatening the realization of this concept is the government's lack of funds.

---

[36]Not that the latter should cause VVS leaders much concern. USAFE's ground attack posture is also but a fraction of what it was at the height of the cold war, and none of the USAF's fighters based in the continental United States are sheltered.

[37]Colonel V. Kistanov, "Composite Air Wings: A New Element in the USAF's Structure," *Zarubezhnoye voennoye obozreniye*, No. 6, 1992, pp. 48–51.

FORCE STRUCTURE AND FORCE DEVELOPMENT PLANS

The baseline for Moscow's planned air strength before the USSR's dissolution was set down in the CFE Treaty signed in November 1990. To comply with the treaty, the Soviet Union was obliged to reduce the number of its combat aircraft west of the Urals to 5150. That ceiling included fighters of Frontal Aviation and VPVO, LRA medium bombers (heavies were regulated by the separate Strategic Arms Reduction Treaty), and combat-capable trainers of the VVS and VPVO. (The L-29 and L-39 basic jet trainers were excluded.)

During the CFE negotiations, the USSR had reported its overall number of combat aircraft west of the Urals to be 6611, including 4323 VVS aircraft and 1338 attack helicopters (225 in the VVS inventory and the remainder assigned to army aviation). After ratification of the treaty, the VVS was to retain 3590 combat aircraft, including 300 medium bombers, with 1550 combat aircraft in VPVO, including 440 combat-capable trainers in the VVS and 60 in VPVO. That would have left the USSR with 1461 surplus combat aircraft divided almost equally between the VVS and VPVO, to be destroyed or used either as static displays, maintenance training aids, or target drones. The plan on the eve of the USSR's collapse was for 290 VVS aircraft to be cut up for scrap, 208 reclassified as unarmed trainers, 160 used for targets, 15 assigned as maintenance trainers, and 36 earmarked for static displays.[38]

These numbers have since been overtaken by events as Russia's continuing military drawdown has brought VVS force levels well below the mandated CFE ceiling. General Deinekin reported VVS plans to scrap 2000 aircraft in 1993 alone, with a view toward ultimately retaining only current-generation aircraft in the operational inventory.[39] He later stated that the VVS inventory would be reduced by

---

[38]Interview with Colonel General Anatoly Malyukov, *Krylia rodiny*, November 1991. The CFE Treaty was a major factor in the breakdown of some former Soviet secrecy rules. General Malyukov said it was now permissible to identify bases, units, and force levels. Also, the commander of VPVO, Colonel General Viktor Prudnikov, summarily declassified the locations of his fighter bases and the names of VPVO unit commanders in 1992. See Irina Pankova, "The Missile Troops Take a Hit—On Their Secrets," *Nezavisimaia gazeta*, August 3–10, 1992.

[39]"Russia To Scrap 2,000 Aircraft," *Flight International*, March 31–April 6, 1993, p. 5.

yet another third in 1994.[40] Colonel General Malyukov indicated that the VVS had ordered "literally a few" Tu-160s, Su-24s, and Su-27s just to keep the military aircraft industry from dying. He said that no further Su-25s or MiG-29s had been ordered, and that the MiG-21, Su-7, and Su-17 had been retired from active service. A similar fate awaits the MiG-23, MiG-27, and Su-17M, as well as bombers built in the 1960s (the Tu-22, Tu-95M and Tu-95K), which will be removed from the active inventory well before the end of this decade.[41]

A persistent problem confronting the VVS is the great diversity of weapons types in its inventory as a result of since-abandoned Soviet acquisition practices. Kokoshin admitted that this costly approach had long been a source of major operations and maintenance headaches. He said that it would take years for the VVS to rid itself of that burden, since it was no easy matter to write off all the dubious "riches" the Soviet military had acquired over several generations of aircraft development as a result of that acquisition overkill. He promised that the defense ministry would in the future radically reduce the number of weapons and equipment types in its inventory, "concentrating on the best items of equipment and saying a resolute *nyet* to those that fail to demonstrate the requisite quality."

It remains too soon to say what the VVS's force structure will look like once the initial round of post-Soviet retrenchment is completed. General Malyukov frankly admitted that he could not offer a firm projection of the VVS's expected strength, since any figures he might give could be rendered invalid at a moment's notice. He indicated that through a combination of radical downsizing and the introduction of latest-generation technologies, the hoped-for combat potential of the VVS would be about 50 percent greater than that of the former Soviet Air Force. He was careful to add, however, that that forecast was based entirely on VVS paper plans and that its realization will depend ultimately on the outlook for state funding.

Government allocations for the procurement of new equipment by the VVS have dropped precipitously since the breakup of the USSR.

---

[40]Interfax, April 13, 1994.

[41]Interview with Piotr Butowski, *Jane's Defense Weekly*, p. 15.

According to the VVS analysis, outlays for the development of new platforms fell 70 percent during the first three years following the end of the cold war. In 1991, aircraft procurement was cut 46 percent for frontal air and 41 percent for bombers and transports. In 1992, procurement of tactical aircraft was only 21 percent of the 1991 level, with that of bombers and transports down to 36 percent of the preceding year's figure. Procurement funds were cut again by 31 percent for frontal air in 1993, with another cut of 23 percent in 1994.

The effects of these cuts can be seen dramatically in the discrepancy between the numbers of military aircraft acquired in 1984 and in 1994, respectively. During the early 1980s, to take a typical annual Soviet figure, the USSR produced an average of 400–450 fighters and 100 bombers and transports a year. By contrast, only 23 fighters were procured by the VVS, VPVO, and Navy combined in 1993–94. By the end of 1994, production of the Su-24, Su-25, and MiG-29 had been terminated; annual production of the MiG-31 for VPVO was down to "single numbers"; procurement of developmental variants of the Su-27 was anticipated to decline to below 14–16 aircraft a year; and production of the Tu-160, Tu-142, and Tu-95 was halted. According to the VVS analysis, state funds allotted to the VVS for procurement in 1993–94 were insufficient even to pay for the replacement of aircraft lost in accidents. The analysis warned that if this trend continues, Russia's military aviation industry could begin to collapse after 1995, causing the country to lose its competitive position as a global aerospace power for decades to come.

On that last score, the VVS analysis grimly concedes that VVS force structure goals mandated by the new military doctrine cannot be met with projected state funding. Merely to sustain an active inventory of 2000 fighter and attack aircraft while replacing the MiG-23, MiG-27, Su-17, and older Su-25s with new equipment, Russia would have to produce 110–115 new aircraft a year to the year 2000. The analysis points out that the Yeltsin government's current and likely future defense budget will not support those production rates. As a result, curtailed procurement will cause the VVS's tactical aircraft inventory to decline from the mandated goal of 2000 aircraft to around 1440 by the year 2000. Once that occurs, even a full economic recovery by 2010 would not enable the VVS to build back up to its mandated level. Furthermore, this excludes the added cost associated with the concurrent need to modernize LRA, VTA, VPVO, and the navy. The analysis concludes that these problems "cannot be solved by any VVS planning or budgeting," but rather are "fundamentally a matter of national priorities and must be addressed at the highest levels of the Ministry of Defense, and by the president, the Security Council, and Federal Assembly of the Russian Federation."

Frontal Aviation

General Antoshkin's recently constituted Frontal Aviation Command (KFA) operates and maintains all tactical aircraft in the VVS inventory. Its establishment withdrew Russian fighter aviation from immediate ownership of the regional military district commanders, long the practice throughout the Soviet era, and reassigned it to the direct control of the VVS. Like the USAF's Air Combat Command, its primary mission is to train and equip Russia's tactical air forces for operational commitment as necessary to joint commands like the Mobility Forces that are now being created.

Like all VVS components, KFA has shrunk considerably from a high of over 5000 combat aircraft in 1989 to less than half that number today (see Figure 3). Now the centerpiece of what has become, in effect, a tactical air force with airlift and bomber backup, KFA today maintains some 2300 combat aircraft. Of these, about a third are fourth-generation MiG-29s and Su-27s. The remainder are older aircraft slated to be retired before the end of this decade.

According to the VVS analysis, KFA in 1994 listed on its roster of assets 450 MiG-29s, 190 Su-27s, 490 Su-24s, 170 Su-25s, and 480 MiG-27s, plus an assortment of MiG-23s, MiG-25s, and Su-17s that are no longer active and are slated for imminent retirement. The command is broken down into five air armies, each made up of three divisions containing three regiments of three squadrons each. A KFA division typically operates 90-120 aircraft depending on type and mission.

KFA also maintains two operational conversion centers with a total of some 300 MiG-29s, Su-24s, Su-25s, and Su-27s for transitioning new undergraduate pilot training (UPT) graduates onto the equipment they will fly on squadron service. In addition, it operates the advanced combat training center at Lipetsk some 400 miles south of Moscow, which develops KFA training and readiness standards, pro-

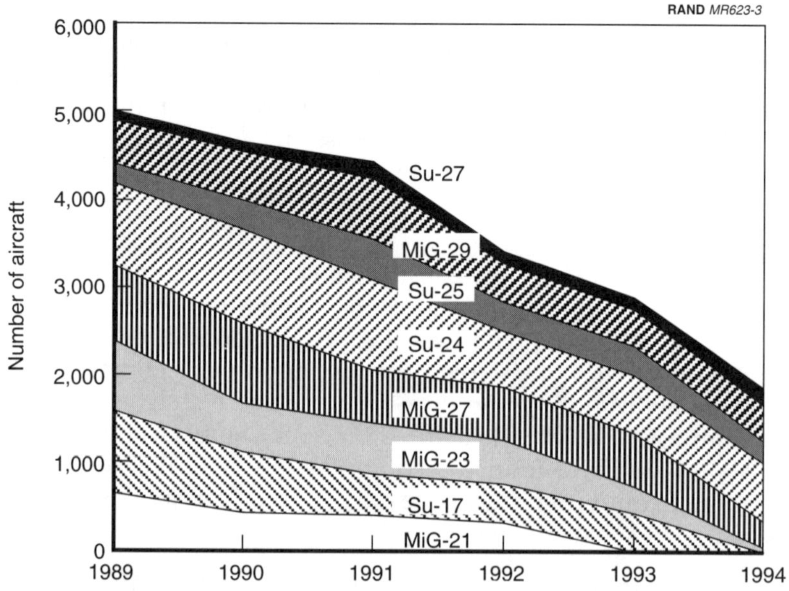

Figure 3—Frontal Aviation, 1989–1994

duces fighter weapons instructors, and offers mission employment training to experienced aircrews. How many of KFA's aircraft assigned to line units and at these training centers are actually in operational service is hard to say, since the fuel crisis has cut annual flying hours to crisis levels and has forced the majority of pilots and aircraft to stand down.

The VVS analysis anticipates that if current budget trends continue, Frontal Aviation's combat aircraft holdings will decline from 2280 to 1670 aircraft in 1997; to 1440 in 2000; to 1330 in 2005; to 1140 in 2010; and to 870 by 2015 (see Figure 4). This projection, moreover, does not reflect anticipated peacetime attrition, currently running at about 1.2 percent a year for fourth-generation aircraft and two percent a year for the remaining aircraft. By the year 2000, according to the analysis, normal attrition could result in a loss of up to 60 third-generation aircraft. By 2015, it could occasion a loss of up to 215 fourth- and fifth-generation aircraft.

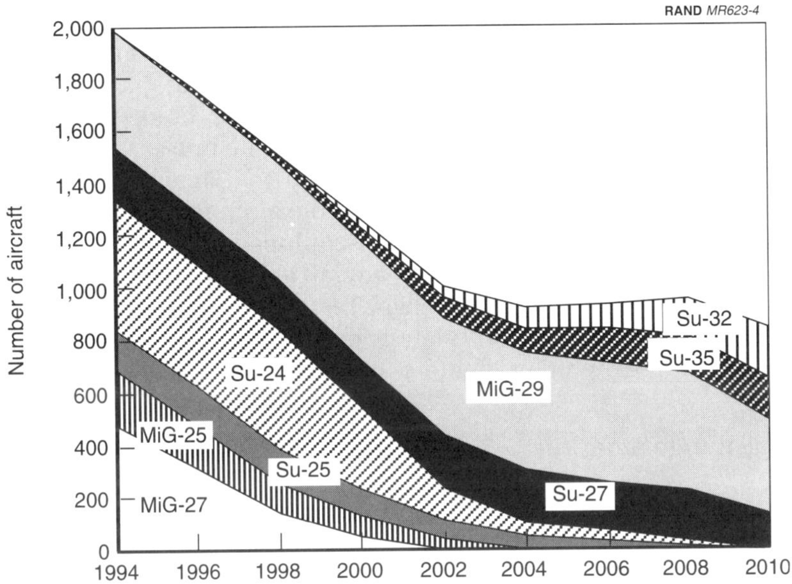

Figure 4—Frontal Aviation to the Year 2010

The outlook for a new Russian air superiority fighter to replace the MiG-29 and Su-27 will be deferred for more detailed treatment in Chapter Nine. It should be noted here, however, that the VVS is facing an uphill struggle even to gain the needed funding to support its desired modernization of *existing* systems, let alone underwrite the acquisition of a next-generation fighter. The VVS analysis anticipates that a new multirole close support fighter (possibly the Su-37) with low-observable features could come on line after the year 2000 to replace the Su-25. It makes no reference whatever in its force projections, however, to any MiG-29 or Su-27 follow-on.

The VVS plans to acquire no more MiG-29s, meaning that the MiG-29M program is dead other than for export, should the Mikoyan Design Bureau and Russia's arms export agency Rosvooruzheniye succeed in finding a foreign market for it.[42] The VVS intends to base its future fighter force on air-to-air, ground attack, and reconnaissance and electronic-countermeasures (ECM) upgrades of the basic Su-27 airframe. The chief of the VVS's operations directorate points out that the Su-27 was designed from the outset with a view toward preplanned product improvement.[43] The two-seat, all-weather ground-attack improvement of the aircraft has been designated Su-32. The single-seat, canard-equipped air-superiority upgrade will be called the Su-35. Both aircraft are now flying in demonstrator form at the Ramenskoye Flight Test Center. Interestingly, and perhaps revealingly, the VVS analysis describes *these* as Russia's impending "fifth-generation" fighters.

Long-Range Aviation

The VVS's bomber component, under the command of Colonel General Igor Kalugin, is Russia's counterpart to the airbreathing portion of the former USAF Strategic Air Command. Once a core component of the Soviet Union's nuclear triad, LRA has shed much, though not all, of its former intercontinental nuclear attack role and replaced it with a new mission: providing extended strategic "reach" in support of Russia's newly emerging regional power projection needs. It also has experienced a significant drawdown in deployed forces since the late 1980s as a result of arms reduction agreements and the USSR's collapse (see Figure 5).

Currently down to a single air army of between two and four bomber divisions, LRA lists on its roster of assets 145 Tu-95s, 120 Tu-22Ms, and four Tu-160s, plus 40-odd Mya-4 and Il-78 tankers and assorted

---

[42]See Mikhail Levin, "That Same MiG," *Krylia Rodiny*, No. 3, 1993, p. 14, and Colonel Valentin Rudenko, "MiG-29M: The Fighter's New Role," *Krasnaia zvezda*, August 20, 1993. A notable deficiency of the otherwise outstanding basic MiG-29 airframe is its limited internal fuel capacity, resulting in limited combat persistence. Russian pilots have described the aircraft jokingly as "a fighter to gain air superiority over the airfield beacon." See Alexander Velovich, "Mikoyan's Market Buster," *Flight International*, September 16–21, 1992, p. 82. Among the MiG-29M's many improvements over the basic MiG-29, the removal of the inlet screens and louvers permitted the addition of 1500 liters of internal fuel with no growth in empty weight.

[43]"The Follow-On to the Su-27 Fighter," paper delivered at the First International Air Power Conference, London, February 11–12, 1993.

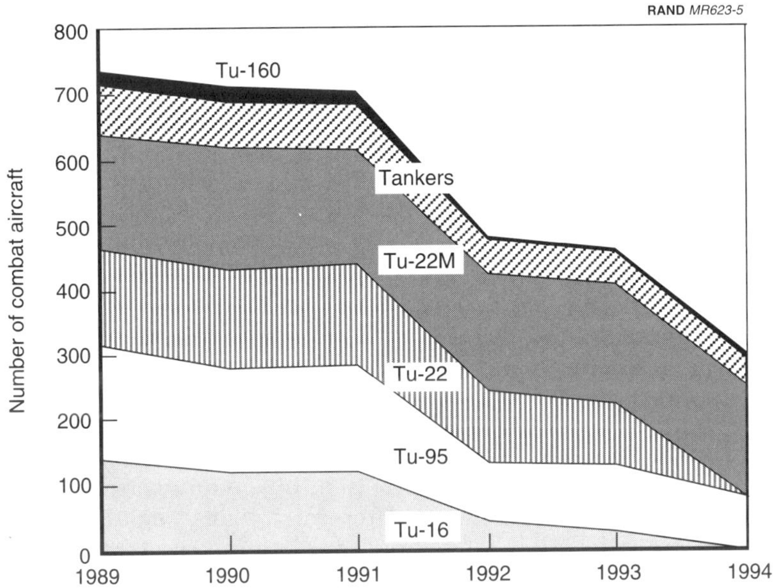

Figure 5—Long-Range Aviation, 1989–1994

transports, for a total inventory of some 400 aircraft. Myriad ownership questions arose in the immediate aftermath of the USSR's collapse, when LRA lost many of its strategic bombers to the newly independent states. More than half of the aviation complement of the former USSR's strategic nuclear forces ended up in Ukraine and Kazakhstan. Of the 160 LRA strategic bombers altogether that had been based on non-Russian Soviet territory, 70 percent were modern Tu-95s and Tu-160s.

The government of Kazakhstan has been quite reasonable with Moscow in agreeing to return those inherited assets for which it has no legitimate use. The last four of a total of 40 late-model Tu-95MS cruise missile carriers were returned to Russia from their base near Semipalatinsk in Kazakhstan in February 1994. Some older Tu-95s were left in Kazakhstan for the Kazakh Air Force.[44]

---

[44]"Strategic Bombers Leave Kazakhstan," *Washington Times*, March 1, 1994.

The story with Ukraine has been something else. The sole operational unit of Soviet Tu-160s, the 184th Heavy Bomber Regiment, was based at Priluki in Ukraine. Upon the USSR's dissolution, these aircraft immediately fell into Ukrainian hands and triggered a testy contretemps between Moscow and Kiev over their return to Russia as a part of the CIS's strategic forces. Only three serviceable Tu-160s were retained in Russia, with a fourth subsequently delivered by the manufacturer to the LRA base at Engels, which had been intended all along to be the VVS's master Tu-160 base upon completion of its construction.[45] In April 1994, production of the Tu-160 at the Tupolev plant in Kazan was reportedly brought to a close by a direct order from President Yeltsin.[46]

For a time, the Ukrainian government was demanding upward of two billion rubles from Russia for each Tu-160 it might return, a price tantamount to extortion, considering that the aircraft were wholly a windfall acquisition by Ukraine.[47] After much tough negotiation on both sides since then, the VVS is about to reacquire ten of the aircraft at a more agreeable and manageable cost.[48] It will be interesting to see what condition the aircraft are in and how much refurbishment they will require by the VVS, since they have been all but nonflyable for five years as a result of inadequate fuel provisions, maintenance, and spare parts.[49]

---

[45]One of those conducted a live cruise missile launch in late 1992. See Sergei Prokopenko, "A Russian Tu-160 on the Combat Line," *Krasnaia zvezda*, October 23, 1992, and Viktor Beltsov, "Russian Air Force Aircraft Inventory Augmented," *Krasnaia zvezda*, February 13, 1993. General Deinekin later indicated that this launch occurred during delivery of the aircraft from the factory to its base.

[46]Moscow television report, April 30, 1994.

[47]ITAR-TASS report, March 23, 1993.

[48]The VVS will also get fifteen Tu-95MS bombers returned. In an offset arrangement, Russia has agreed to transfer eight MiG-29UB two-seaters, four Su-27s, and several transport aircraft to Ukraine. See "Ukrainian Bombers for Russia," *ConCISe*, April 23, 1996, p. 461.

[49]As a testament to the Tu-160's problems, the report added that the first aircraft was accepted at Priluki from the factory in April 1987 and since then only nineteen Tu-160s had been delivered altogether. The report added that ground servicing of this modern aircraft was "prehistoric," requiring ten fuel trucks to transfer its full complement of 180 tons of fuel. One aircraft was lost to an engine fire due to an electrical system problem. All four crew members ejected successfully. See Leonid Kostrov, "The Blackjack: Not a Game for Politicians," *Vecherniaia Moskva*, March 3, 1992.

At much the same rate as KFA, LRA will experience a sustained drawdown through the first decade of the 21st century from its current posture of somewhat more than 400 bombers and tankers to an inventory roughly half that size (see Figure 6). Declared plans are to retire all Tu-16, Tu-22, Tu-95M, and Tu-95K aircraft, leaving a bomber force made up of Tu-22M, Tu-95MS, and Tu-160 aircraft, supported by Il-78 tankers.

Military Transport Aviation

By far the most painful aircraft loss suffered by Russia in the wake of the USSR's disintegration was felt in Military Transport Aviation (VTA), under the command of Colonel General Vyacheslav Yefanov. The largest group of its late-model Il-76 jet transports, and those with the longest remaining service life (200 out of the 450 possessed by the

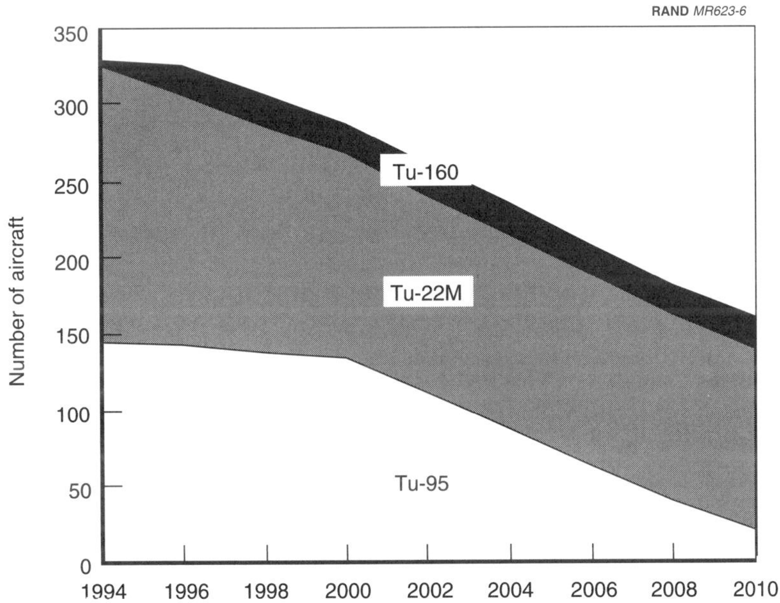

Figure 6—Long-Range Aviation to the Year 2010

USSR), were based in Ukraine.[50] Their loss was doubly disturbing in light of Russia's newly emergent regional peacekeeping challenges and the consequent power projection needs they have engendered, an operational challenge that was sorely tested by the airlift demands of the war in Chechnya three years later.

General Malyukov, a career fighter pilot, freely admitted in 1993 that the VVS's single biggest shortfall was in the realm of air transport. The VVS analysis likewise observed that mobility was the highest priority for the Russian armed forces, adding that the missions assigned to VTA by Russia's new military doctrine "far exceed present capabilities." By way of example, the analysis noted that it would take all of VTA to move just a single airborne division in two sorties and that its capabilities will have to be increased at least threefold if it is to provide the mobility envisaged by the new doctrine.

VTA currently consists of three air divisions, each made up of three regiments possessing some 30 transport aircraft apiece. There are also several independent airlift regiments assigned to the command. The combined inventory includes some 250 Il-76s, 70 obsolescent An-22 turboprop transports, and 20 An-124 heavy airlifters (see Figure 7). The An-12 has been phased out. To triple its existing lift capacity as recommended by the VVS analysis, VTA would have to produce 28 to 30 new Il-106 transports after the year 2000, when this follow-on aircraft is expected to be ready for production. According to the VVS analysis, however, achievement of that goal would completely deprive the VVS of any funds to modernize its other force components.

A serious problem for VTA is that most of its aircraft production facilities were located outside of Russia. For example, the plants that produced the Il-76 and An-124, along with those that made their en-

---

[50]TASS report on a statement by General Deinekin, March 27, 1992. Defense minister Grachev later painted an even starker picture: "We have been left with less than one-half of [the former Soviet] Military Transport Aviation. . . . Two of its five divisions have remained on Ukrainian territory, one regiment has remained in Belarus, and one in Uzbekistan. To all intents and purposes, Russia actually has only two military transport regiments." Interview by Nikolai Burbyga and Albert Plutnik, "Pavel Grachev: 'In Essence the Russian Military Must Be Recreated from Scratch,'" *Izvestiia*, June 2, 1992.

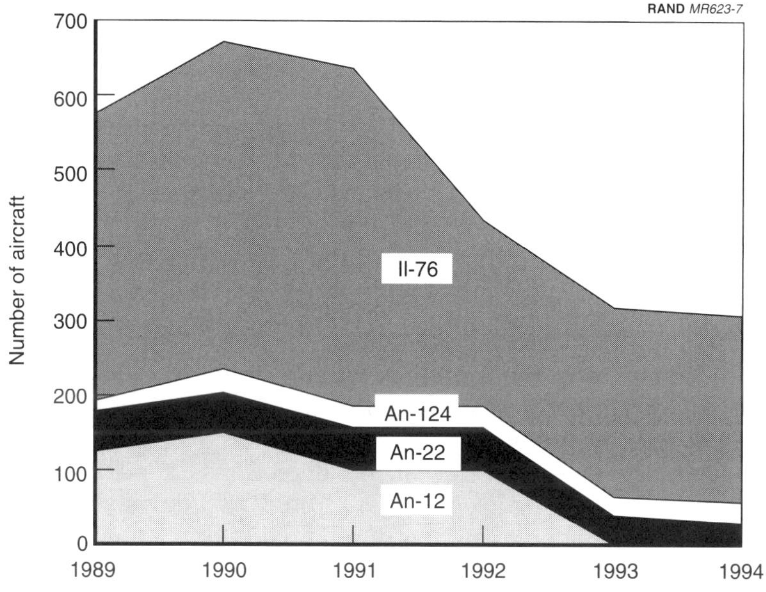

Figure 7—Military Transport Aviation, 1989–1994

gines, were lost to Ukraine and Uzbekistan. To VTA's distress, production of the workhorse Il-76 had to be halted, since the aircraft was assembled in Tashkent, now the capital of Uzbekistan. To take up the slack, the new An-70 medium transport will soon become available. The Antonov factory that produces it, however, is in Kiev, meaning that the aircraft will have to be purchased by the Russian government at market price from Ukraine. As in the case of the Tu-160 buyback, negotiations are now under way for an arrangement that will work to the best interests of both sides.

Reserve and Training Command

The new Reserve and Training Command under Colonel General Leonid Stepanyuk was set up, among other things, to manage the VVS's four undergraduate pilot training schools and several additional schools for navigators, air traffic controllers, and foreign students. It will also maintain the VVS's "boneyard," where surplus aircraft will be held in storage for possible return to service in case of a national emergency. Finally, it will develop and conduct periodic high-intensity courses for recurrency training of VVS reserve aircrews, both in peacetime and on an as-needed basis in case of a crisis requiring mobilization of reserve forces.

Current holdings in the command's inventory include around 1000 L-39 basic jet trainers distributed among the Kacha, Barnaul, Tambov, and Balashov flight schools, as well as some 300 multi-engine Tu-134 jet transports. The command also maintains a thousand or more recently retired combat aircraft of various types (mainly MiG-23s, MiG-27s, and Su-17s) sequestered at storage bases.

General Stepanyuk's organization is planning to form a "first-line reserve" in the near future made up of units withdrawn from the Warsaw Pact countries and former republics and of obsolescent aircraft placed in flyable storage from decommissioned units. A "second-line" reserve, according to the VVS analysis, will be equipped with older combat aircraft retained in nonflyable storage. The latter will require bases configured for the long-term preservation of stored aircraft.

The command anticipates that once the second-line reserve system is established and activated, it will be possible to reconstitute enough aircraft for a single fighter regiment in the space of about a week. A serious problem is that many of the VVS's stored aircraft are rapidly deteriorating beyond the point of further serviceability as a result of exposure to harsh elements. Another is that vital components and subsystems of many stored aircraft are being stolen on a massive scale because of inadequate security and accounting.

The VVS badly needs a new basic trainer and has a stated requirement for 800 aircraft.[51] Its aging L-39s have become a maintenance nightmare. Manufactured in Czechoslovakia and procured in large numbers by the Soviet Air Force as a "burden-sharing" gesture to its then–Warsaw Pact ally, the aircraft are now said to be "catastrophically" short of spare parts. Requests for replacement parts go routinely unfulfilled. Supplies in hand are said to be so scarce that, according to one complaint, "you can't find the simplest 3.5 volt light

[51]Igor Andreyev, "Aircraft Competition: MiG or Yak?" *Izvestiia*, February 22, 1994.

bulb in an entire regiment, even in daylight and with a flashlight."[52] Currently in competition to replace the Czech-built L-39 are Mikoyan's MiG-AT and Yakovlev's Yak-130, with a VVS source selection and production decision still pending after more than two years of being reported as "imminent."[53] Both aircraft were on static display at the 1995 Paris and Moscow air shows and since have made their initial test flights.[54]

Air Defense

Russia's Air Defense Force (VPVO), under the command of Colonel General Viktor Prudnikov, remains a separate service which operates all Russian homeland surface-to-air missiles, early warning radars, ground-controlled intercept (GCI) sites, and fighter-interceptors. The fighter aviation component of the force, IA/PVO, is commanded by Colonel General Vladimir Andreyev, a highly regarded fighter pilot who, among other things, conducted the first in-flight refueling of a MiG-31 in early 1994.

IA/PVO's inventory has experienced a rate of decline much like that of the other combat air arms since the late 1980s. From a high of some 2300 operational interceptors on the eve of the USSR's collapse in December 1991, it is down to less than half that number today (see Figure 8). Organized into five air defense armies, mostly concentrated in Western Russia and the Far East, the command currently operates some 325 Su-27s, 425 MiG-31s, 20 Il-76 AWACS aircraft, and a residual number of MiG-23s awaiting retirement. It also maintains a UPT school at Armavir, which operates some 225 L-39s and 73 MiG-23s, a similar flight school at Stavropol which also trains

---

[52]Major V. Malev, "A Problem Requiring a Decision," *Krasnaia zvezda*, September 21, 1991.

[53]See Alexander Velovich, "Russian Trainer Decision Close," *Flight International*, February 22–27, 1994, p. 22.

[54]See "MiG-AT Trainer Completes First Flight," *Aviation Week and Space Technology*, April 11, 1996, p. 21, and "First Flight for Yak-130 Trainer," *Aviation Week and Space Technology*, May 6, 1996, p. 17. As one option that seems poorly advised from an operational and cost perspective but designed to help keep both Mikoyan and Yakovlev from going under, the VVS is said to be considering a mixed buy of both aircraft.

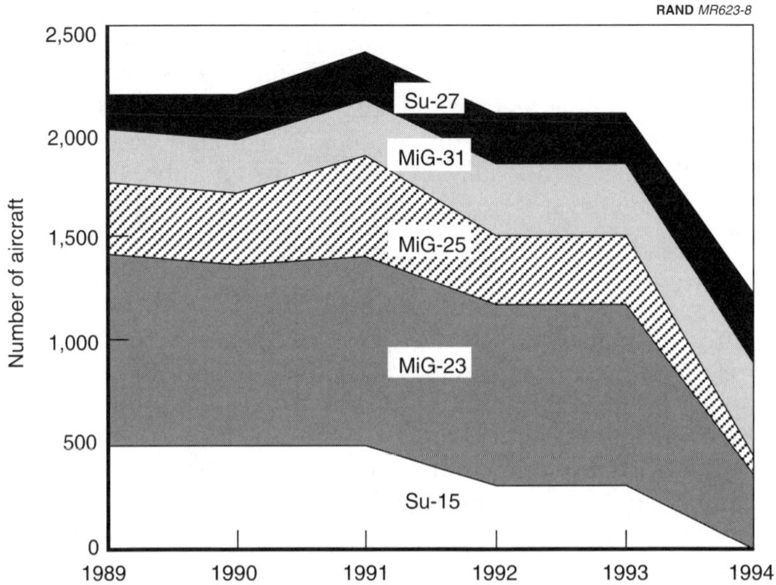

Figure 8—Air Defense Forces, 1989–1994

MiG-31 radar intercept operators, and an advanced interceptor weapons training center at Savostleika.

The VVS analysis points out that if IA/PVO eventually merges with the VVS—a move now said to be under "intensive debate"—these assets would become an important part of the VVS's overall force posture. They would also significantly affect its existing operational planning and procurement needs. Force modernization plans appear limited to developing and producing an advanced MiG-31M if and when procurement allocations permit. Budget constraints promise a further decline in VPVO fighter assets over time, with the current force of some 1100 interceptors expected to fall to as low as 380 by 2010 (see Figure 9).

Naval Aviation

Although they belong to a separate service, Russia's naval aviators wear VVS uniforms, receive their training at VVS flight schools, and

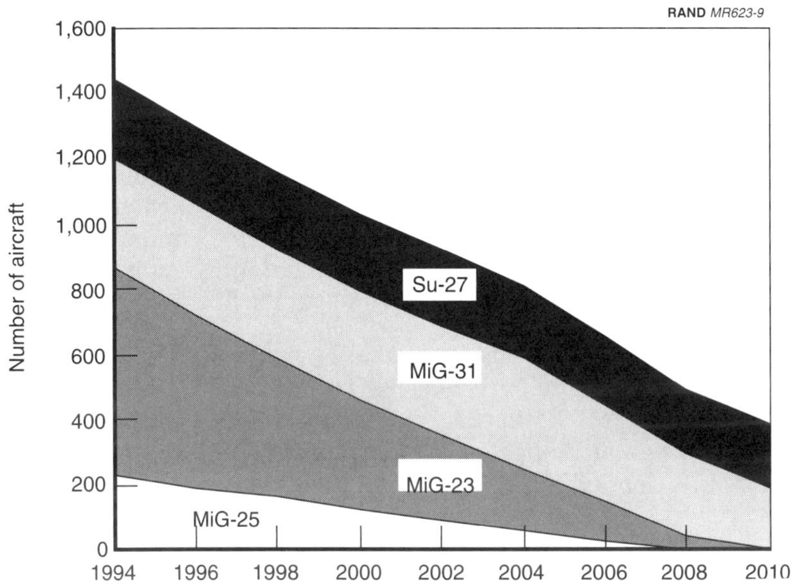

Figure 9—Air Defense Forces to the Year 2010

hold VVS rather than naval ranks.[55] They also are afflicted by post-Soviet constraints no less severe than those that have beset the VVS and VPVO. During its final days, the Soviet Navy maintained over 1000 aircraft altogether, consisting of shore-based fighters and fighter-bombers, cruise missile–carrying bombers, maritime patrol and reconnaissance aircraft, anti-submarine warfare (ASW) helicopters, and transports. That number has since declined dramatically in consonance with Russia's post–cold war military drawdown (see Figure 10), aside from a transitory exception in 1991 when the navy received a transfer of 290 "excess" Su-24, Su-25, MiG-27, and Su-17 fighter-bombers from the VVS. This was probably part of a Soviet ploy to help the VVS evade CFE Treaty limits.

According to its first deputy commander, Major General Nikolai Rogov, naval aviation during the Soviet era was responsible for pro-

---

[55]See Captain Brian J. Collins, USAF, "Soviet Military Reform: Structuring Aerospace Forces," *International Defense Review*, June 1991, p. 562.

tecting nuclear ballistic missile submarines (SSBNs) and submarine cruise missile platforms, engaging enemy surface action groups, interdicting sea lines of communication, and aiding in maritime defensive and offensive operations. These missions have since been scaled back from their former blue-water orientation to a continental focus, in keeping with the across-the-board retrenchment of Russia's military power and reach. General Rogov has acknowledged a 30-percent reduction in naval personnel and a 20-percent drawdown in the navy's aircraft inventory, with priority attention now being devoted to the air arms of the Northern and Pacific fleets.

Among the many proposals for further consolidating Russia's radically trimmed-down military, the idea has been raised to subordinate the navy's Tu-22M bombers to VVS operational jurisdiction and to transfer its fighter and ground-attack aircraft directly to the VVS. Predictably, General Rogov has voiced exception to this idea, on the

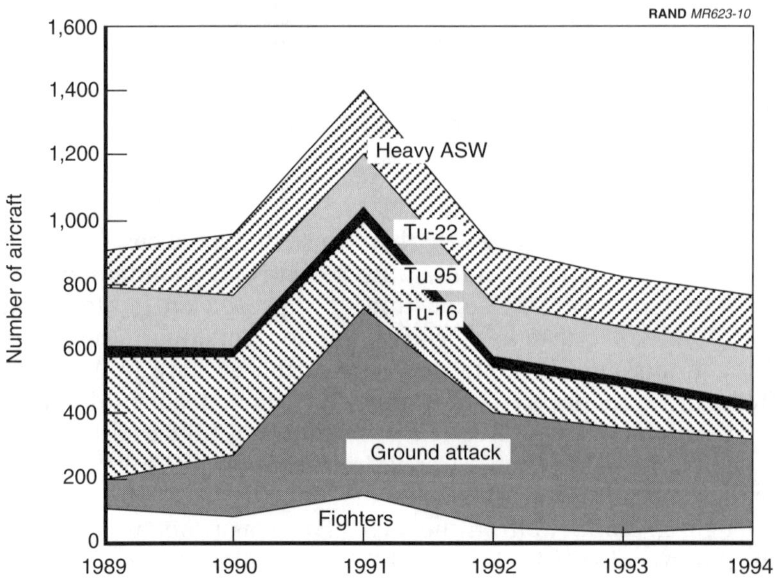

Figure 10—Naval Aviation, 1989–1994

ground that its implementation would, in his words, "deprive the navy of organic strike aviation to counter the threat of air and missile attack."

Were the navy's combat aircraft to be taken over by the VVS, General Rogov maintained, the fleet would then have to rely on VVS units untrained for air war at sea. Such an arrangement, he feared, would create new command and control problems. To the contrary, he suggested, the continuing decline in Russia's submarine and surface warfare strength demands an offsetting "reinforcement" of naval aviation. Absent a compelling strategic rationale, this has an all too familiar ring of rote justification of existing roles and missions. At the least, it promises to make for some lively bureaucratic swordplay between the navy and the VVS as Russia's force drawdown continues.

Perhaps anticipating worse things to come, General Rogov took care not to reject altogether the idea of combining LRA and the navy's cruise-missile aviation. He suggested, however, that considering the predominance of naval air's responsibilities in the maritime arena, it made more sense to subordinate it to the navy. He also found it more sensible, in light of Russia's diminished post–cold war maritime threat, to retain the long-range missile aircraft of the Northern and Pacific fleets but to reduce the navy's involvement in fighter-bomber, ASW, and reconnaissance aviation.[56]

The navy's commander in chief, Admiral Feliks Gromov, has reported that naval aviation will be reduced further by approximately 40 percent, with chief emphasis being given to the development of carrier-based fighters, cruise-missile carriers, and ASW aircraft. He has frankly admitted, however, that because of the continuing funding crisis, there will be "serious difficulties in the practical attainment of these ideas."[57]

A more upbeat assessment of naval aviation's durability in the face of the current crisis was offered by its commander, Colonel General Potapov, who stressed that even with reduced funding, fuel shortages, and maintenance deficiencies, continuation training in operational units remained aimed at preserving mission readiness among enough aircrews whereby, "if necessary, every flying unit would be

[56]Interview with Major General of Aviation Nikolai A. Rogov, "Russia's Naval Aviation: What Kind Should It Be?" *Morskoi sbornik*, No. 8–9, August–September 1992, pp. 3–5.

[57]Admiral F. Gromov, "The Russian Navy: Yesterday, Today, and Tomorrow," *Morskoi sbornik*, No. 1, January 1993, p. 7.

able to carry out its primary mission."[58] General Potapov acknowledged that dwindling aircraft service life was threatening to ground a considerable portion of the aircraft inventory. He added that the navy intends to press ahead in ASW, leaving unaddressed the hard question of where Russia's requirement for this capability remains in the rank-ordering of military needs in today's post–cold war world.

Much as in the VVS, there has been a significant rise in naval aviation's accident rate as a result of the prolonged disruption of shipboard helicopter and aircraft operations. One aviation colonel complained several years ago that his pilots had not taken part in shipboard operations for more than a year. This, he said, "jeopardizes not just the regiment's combat readiness but flight safety as well."[59]

The navy lost all its aviation training centers to Ukraine following the USSR's collapse, prompting a need to establish at least one flight training center on Russian soil. Without one, the navy has no means of converting its new graduates from VVS flight schools. In 1991, General Deinekin and then–defense minister Shaposhnikov agreed to give up the VVS's flight school at Yeisk to the navy. That agreement was suspended, however, after Grachev assumed office as Minister of Defense. For the moment, Yeisk remains a VVS air traffic control training center. General Potapov insists that the navy needs only a portion of the VVS's academic facilities at Yeisk, plus permission for joint basing for some carrier-based aircraft at the Yeisk airfield for operational conversion training of new aircrews.

General Rogov spoke optimistically in 1992 about the Russian Navy's intent to stay involved in carrier-based fighter aviation. The Soviet Navy's development of the first-generation ASW cruisers *Moskva* and *Kiev* during the late 1960s, and its later development of the second-generation heavy carrier *Admiral Kuznetsov*, hardly gave it a serious ship-based attack aviation capability. Nevertheless, it provided a feel for big-deck carrier aviation that could not have been acquired any

---

[58]Colonel General V. Potapov, "Naval Aviation: Problems and Solutions," *Morskoi sbornik*, No. 6, June 1993, p. 29.

[59]Vladimir Shirokov, Valery Gromak, Aleksandr Pilipchuk, and Valery Polikarpov, "The Deck-Based Pilots Are Losing Their Deck But Believe Their Problems Can Be Solved," *Krasnaia zvezda*, June 30, 1993.

other way. Unfortunately for the navy's ambitions to continue along this route, its initial conversance with the complex demands of carrier aviation was acquired just as the funding crisis threatened the demise of carrier-based air power in Russia altogether.[60] One retired admiral and former deputy chief of the General Staff, Admiral Nikolai Amelko, has even charged that the building of Russian carriers was a "wildcat scheme" all along and totally at odds with the USSR's defensive military doctrine. He complained that aircraft carriers have nothing to contribute to the nation's defenses and merely "eat up resources."[61]

In 1993, the navy selected the Su-27K over the MiG-29K as its preferred shipboard fighter, a choice that resulted in cancellation of further development of the MiG-29K. The carrier version of the Su-25 ground-attack aircraft also appears to have been terminated.[62] Upon becoming operational, the Su-27K was redesignated the Su-33.[63] The carrier *Admiral Kuznetsov* has now deployed with the Northern Fleet and sports a squadron of sixteen Su-33s, eight Su-25s, and ten Ka-27 helicopters.[64] In August 1993, no serving navy pilots had fully qualified for Su-33 carrier operations, although their training was in progress.[65] In the meantime, the navy has phased out its inventory of plagued Yak-38 VTOL fighters.[66] It is unlikely that a

---

[60]Captain First Rank V. Kuzin, "Aircraft-Carrying Cruisers," *Morskoi sbornik*, No. 7, July 1991, p. 64.

[61]Cited in "Does the Soviet Navy Need Aircraft Carriers?" Novosti report, IAN in English, September 6, 1991.

[62]"Sukhoi Beats Mikoyan in Navy Contest," *Flight International*, September 8–14, 1993, p. 11.

[63]Pavel Maslov and Valentin Rudenko, "The Su-33 Attacks from the Sea," *Krasnaia zvezda*, March 11, 1994.

[64]The first-generation ASW carriers *Minsk* and *Novorossiisk* in the Pacific Fleet and the *Kiev* in the Northern Fleet have been retired.

[65]Conversation with Sukhoi Design Bureau test pilots at the Ramenskoye Flight Test Center, August 31, 1993.

[66]The Yak-38 experienced an unacceptably high attrition rate due to reliability problems with its stability augmentation system. A Soviet press account noted that pilots on one carrier hung a sign on a Yak-38 declaring that the airplane was "unfit for human use." See Norman Friedman, "World Naval Developments," *U.S. Naval Institute Proceedings*, September 1990, p. 139. The Yak-141, originally intended to replace the Yak-38, has also been cancelled as a result of development problems, budget constraints, and a questionable post–cold war requirement for the aircraft.

second *Admiral Kuznetsov*–class carrier, the *Riga,* will be finished. The big-deck carrier that had long been under construction at the Nikolayev shipyard has been cancelled. All of this adds up to a grim outlook for Russian carrier-based fighter aviation.[67]

---

[67]See "Russia Ditches Forgers and Carriers," *Flight International,* March 1, 1994.

Chapter Four

UNDERGRADUATE PILOT TRAINING

Throughout most of the cold war, the VVS operated a specialized undergraduate pilot training (UPT) program consisting of a dozen Higher Military Aviation Schools for Pilots (*Vyshiye voenniye aviatsionniye uchilishchiya letchikov*, or VVAULs) located in various Soviet republics. Each school trained cadets for conversion to a specific aircraft category, with curricula tailored variously toward fighters, ground attack aircraft, strategic bombers, and transports, depending on the school. The course covered four years and combined training in officership, basic college-level science and engineering, and entry-level aeronautical skills. Application to a VVAUL meant signing up for a 25-year service commitment in the event of successful completion, with service time commencing the year of enrollment. Graduates of the program at each school were commissioned with the rank of lieutenant, awarded a degree in engineering, and given an aeronautical rating of basic pilot.

Although the VVS and VPVO were and remain separate services, their VVAULs offered virtually indistinguishable training programs. Captain Alexander Zuyev, the VVS MiG-29 pilot who defected to the United States in July 1989, received his flight training at Armavir, a VPVO school, and subsequently was assigned to the VVS.[1] Fighter

---

[1]VPVO and VVS fighter pilots, however, work in markedly dissimilar operating environments, with the former far more heavily slaved to GCI close control throughout intercept missions. Zuyev implied that Frontal Aviation pilots looked upon their VPVO brethren with some condescension, saying of one VVS MiG-29 squadron commander whose upbringing had been on the clumsy Yak-28 in VPVO that "he might well have flown thousands of 'elevator ride' sorties as a VPVO interceptor pilot, but that fact had

schools used the Czech-produced L-29 and, later, L-39 jet trainers for primary and basic instruction, after which cadets would receive an initial qualification checkout on the aircraft they would fly on their first operational assignment (generally the MiG-21 or MiG-23 for air-to-air pilots and the Su-17 or MiG-27 for those headed for ground attack units). This transition took place during the final year of the VVAUL course.

Acceptance to fighter VVAULs was highly competitive. Lieutenant Viktor Belenko, the VPVO MiG-25 pilot who defected with his aircraft to Japan in 1976, stated that of some 4000 applicants for slots in his pilot training class at Armavir, only 360 were selected, and that about a hundred of these were washed out before completion of the program.[2] The most popular schools were more difficult to get into than others, of course. For example, Zuyev reported that the Kacha VVAUL for fighter pilots was all but inaccessible without inside connections or other pull. Zuyev added that in the year he applied, 100,000 tenth-year high school aspirants nationwide had applied for all twelve VVS and VPVO flight schools, with as many as 20,000 applying to Armavir alone. Only about 2000 of the latter passed the rigorous physical and aptitude tests, with 300 ultimately selected for the class beginning in September 1978.

SCREENING AND SELECTION

Because of the Soviet military's pervasive secrecy, little was published about the VVAUL system or the nature of a flying career in the VVS, apart from pamphlets that mainly played up the romance and patriotic appeal of military aviation.[3] As a result, in marked contrast to American and British pilot recruiting practice, most information for interested Soviet youths was passed along by word of mouth, with family acquaintances from the retired pilot community often playing an important part.

---

little to do with his true ability as a fighter pilot." Alexander Zuyev, with Malcolm McConnell, *Fulcrum*, New York, Warner Books, Inc., 1992, p. 210.

[2]John Barron, *MiG Pilot: The Final Escape of Lieutenant Belenko*, New York, McGraw-Hill Book Company, 1980, p. 59.

[3]For a classic example, see Lieutenant Colonel V. Smirnov, *Vybiraiu aviatsiiu: Dlia tekh, kto khochet stat' ofitserom VVS* ("I Choose Aviation: For Those Who Want to Become an Air Force Officer"), Moscow, Voenizdat, 1989.

Applications were solicited from civilian high school students, serving conscripts and career noncommissioned officers, reservists who had completed their compulsory service obligation, active-duty warrant officers, military construction personnel, and secondary school cadets in the Suvorov military and Nakhimov naval boarding schools.[4] The enrollment age window for civilian youths was 17–21 years, with a 23-year age cap for all other applicants. Interested high school students obtained information on available flight schools and application procedures from their town or regional military commissariat (*voenkomat*), the Soviet equivalent of an American draft board. Military personnel were directed to apply through their unit commander by April 1 of the desired year of entry. Civilians could apply to their *voenkomat* up to May 1. In all cases, application was to a specifically requested VVAUL.

Initial screening tests were administered by the applicant's local *voenkomat*. These were followed by a series of three increasingly exacting medical examinations. Candidates were also screened for political rectitude and reliability, with special concern about religious family members or any relatives who had traveled abroad. The successful applicant had to produce a strong character and political suitability recommendation from his DOSAAF or high school military instructor.

Voenkomat staffs would complete their prescreening by May 15 and forward their recommendations to the VVAULs by June 5. By June 30, the VVAULs would inform those applicants who had been selected when to arrive at the school for further testing. Admission of non-Russian or non-Slav candidates into either VVS or VPVO VVAULs was the exception to the rule, although such candidates often became outstanding and respected pilots.[5]

---

[4]Lieutenant Colonel V. Smirnov, "For Those Who Choose the Sky," *Aviatsiia i kosmonavtika*, No. 3, March 1990, pp. 40–41.

[5]One notable example being Tokhtar Aubakirov, who while deputy chief test pilot of the Mikoyan Design Bureau made the first arrested landing of a MiG-29 aboard the Soviet carrier *Admiral Kuznetzov* in 1989 and later went on to fly in space as a pilot-cosmonaut. Others include the late Major General Sulambek Oskanov, the respected commander of the VVS's fighter weapons training center at Lipetsk, who was killed in a MiG-29 accident caused by an uncommanded roll during approach to landing in 1992.

Those surviving the first cut proceeded by train to the VVAUL to which they had applied, where they lived under tents for four weeks of further physical, psychological, and academic testing. Throughout the evaluation and screening phase, all applicants were treated as conscripts and served standard conscript meals. Between examinations, they would do routine conscript work, such as unloading bricks from flatbeds, digging ditches, weeding fields, and laying concrete slabs for runways. During this time, they were closely observed and rank-ordered by the school staff.[6]

Medical standards were uncompromising, with qualifying applicants expected to meet the cosmonaut criteria of a Category 1 health certificate. In practice, most line MiG-29 and Su-27 pilots maintained a Category 2 certificate, which was only slightly less demanding. However, they had to maintain 20/20 vision (called "one-by-one" in the Soviet system). No pilots who wore glasses flew fighters in the VVS.[7]

The VVAUL entry written exam covered secondary school subjects, including mathematics and physics, Russian language and composition, and the history of the USSR. A special waiver was allowed for graduates of the Suvorov and Nakhimov military boarding schools and in the rare case of a military applicant who had received the Hero of the Soviet Union award for valor. These were accepted without testing in general subjects, assuming that all other standards were successfully met. The demanding mathematics and physics tests weeded out applicants from rural areas whose high school preparation was substandard.

The physical fitness test included bar chin-ups, a timed 100-meter dash, a 3000-meter cross country run, and a 100-meter swim. The psychological exam consisted of a diagnostic questionnaire and other analytic tests. Applicants were also screened for motor skills,

---

[6]The Soviet approach to pilot selection showed many similarities to that employed by the heavily fighter-oriented Israeli Air Force. For early details on the latter, see Edward W. Youngling et al., *Feasibility Study to Predict Combat Effectiveness for Selected Military Roles: Fighter Pilot Effectiveness*, St. Louis, McDonnell Douglas Astronautics Company, MDC E1634, April 29, 1977, pp. 3.79–3.83.

[7]Zuyev, *Fulcrum*, p. 123.

with the hand-eye coordination test being administered by means of a crude control stick and "gunsight" apparatus.

The evaluation process divided the candidates into three groups, ranging from qualified (Group One) through borderline (Group Two) to unqualified (Group Three). Those who made the final cut entered the fall course as beginning fourth-class cadets (*kursanty*). Special amenities promised to cadets as aspiring career aviators included free meals seven days a week, 45 days of paid leave annually, travel anywhere in the USSR on Aeroflot for a modest fee, and retirement at age 40 with two-thirds pay for the rest of one's life. The best medical care in the USSR was also available, along with enormous social prestige and other privileges.[8]

THE UPT SYLLABUS

At Armavir in 1978, according to Zuyev, the 1100 cadets who made up the school's complement of pilot trainees were divided into four sections. The school conducted an instruction plan that had half the cadets in all classes at any given time on deployment to outlying airfields for flight instruction on the L-29 or MiG-21, and the other half remaining at the school for academics or additional flying. Auxiliary training fields were remote from the school, often as far as 100 miles away.

During the first year, the daily schedule began with an 0600 wake-up and 45 seconds to dress for running and calisthenics, followed by wash-up and preparation of the barracks for inspection. Academic instruction began at 0730 and continued for a 12-hour day to 1930 six days a week, with lights out at 2200. Sunday was not free but involved cleanup duties, as well as mandatory lectures by the school's political officer (*zampolit*). Zuyev remarked that the cadets hated these and would often slip a flight manual inside their Marxist-Leninist texts to gain some value from what they otherwise considered a waste of time.[9]

---

[8]Ibid., p. 123.

[9]Ibid., p. 125.

Evidently there was little or no systematic hazing. Zuyev noted that VVAUL *kursanty* were aware that first-year cadets at American military academies were regularly harassed by upperclassmen. He added that "we found that a totally alien concept. Here we needed mutual support to survive the rigorous work load."[10] However, there was an abundance of snitches (*stukachi*) among one's classmates and the staff who would readily tattle on one to the school's commander or, worse yet, to its KGB "special section" (*osobiy otdel*) for any observed political indiscretions or other malfeasance. Cadets had to be wary at all times against these omnipresent spies.

Cadets were taught that American pilots were able and courageous and would stay and fight rather than turn and run if confronted in the air. They were also briefed that the USAF used special psychological screening tests to identify and single out applicants with the greatest aptitude for combat flying.[11] Cadets were given security clearances and allowed to read periodic intelligence reports on the American and NATO air threat. These were said to be straightforward, objective, and devoid of sermonizing.

Parachute training was mandatory. This involved initial qualifying jumps out of an An-2 and a minimum of two static-line jumps annually thereafter. Remarkably, the Soviets did not use altitude chambers to instruct aircrews to recognize their hypoxia symptoms, either in VVAULs or in operational units, despite their awareness that foreign air forces employed them routinely for high-altitude indoctrination. The VVS does operate altitude chambers for aeromedical research and, on occasion, to test aircrews for hypoxia susceptibility. But to this day, it does not conduct a regular physiological training

---

[10]Ibid., p. 85.

[11]Barron, *MiG Pilot*, p. 66. In fact, it did anything but, at least through the mid-1980s. By one informed account, Soviet competition for pilot training slots was more competitive than that of the USAF, which did "very little preselection testing of personnel prior to their entering pilot training," relied on "a relatively antiquated system of undergraduate academic grades, officer qualification test scores, and 20/20 vision," and did not "differentiate between skills necessary for fighter pilots and other pilots such as airlift or bomber pilots." In one weakness the USAF shared with the Soviet VVAUL system, such a distinction was made much later in the training cycle and enabled the system to select only from among those who had already been admitted into the program. Colonel Mike Press, "The Human Factor: The United States Versus the Soviet Fighter Pilot," *Air University Review*, November–December 1986, p. 74.

program, even though aircrews and medical personnel have more than once called for it.[12] There is also no routine provision for aircrew *g*-awareness orientation or *g*-tolerance training on a centrifuge.[13]

Preflight instruction on the L-29 or L-39 began at the end of the second semester of the first year of academics, with six months of flying commencing at the start of the second year after a summer vacation break. During the flight training phase, wake-up was at 0400 for an early breakfast, followed by as many as three sorties a day, with additional classroom academics in the afternoon. Short sorties were the rule, and it was common for cadets not to be cleared for solo flight until after 36 or more dual instruction rides. The L-29/L-39 primary phase included contact flying, navigation, formation, and basic aircraft handling and aerobatics, with each training flight capped by a return to base on the automatic direction-finding (ADF) needle. Toward the end of basic training, cadets participated in a judged maneuver competition. The last year of the VVAUL program was devoted entirely to flying.

The flying program at Armavir during the early 1970s included 100 hours on the L-29 and 200 hours on the MiG-17, for a total of 300 dual and solo hours. A decade later, Zuyev completed the program with only 230 hours, all on the L-29. He was one of fifty top-graduating cadets in his class who converted from the L-29 directly to the MiG-23. Other cadets upgraded first to the MiG-21. The conversion syllabus began with an extensive ground school on aircraft systems and procedures, followed by initial flights in a two-seater. Later, as the L-29 was phased out and replaced by the more modern L-39, primary flight training in the VVS consisted of sixty hours and basic training an additional ninety hours, the latter concentrating on air-to-air maneuvering, air-to-ground gunnery, two-ship tactical formation, and squadron-level operations. In 1985, a new VVAUL landing-instruction syllabus incorporating both straight-in and 360-degree overhead approaches was issued. This

---

[12]Colonel of Medical Service (Ret.) I. Chernyakov and Lieutenant Colonel A. Shishov, "Diagnosing Hypoxia in Flight," *Aviatsiia i kosmonavtika*, No. 11, November 1991, pp. 10–11.

[13]Conversation with a former Soviet fighter pilot.

reportedly resulted in fewer hard landings, better speed control on the base turn, and elimination of ten dual flights from the syllabus.

OPERATIONAL CONVERSION

Through the mid-1980s, the VVS provided its prospective fighter pilots with initial qualification training on their assigned operational aircraft during the final year of the VVAUL course. Cadets graduated as basic, nonrated pilots cleared to fly the MiG-21 or MiG-23 only within a narrow band of operations, to include clear-weather takeoffs and landings, formation flight, and basic ground attack and aerial maneuvering. They would then report to their assigned unit for upgrade training to mission readiness. Zuyev reported that "agents" from line regiments would visit Armavir and other VVAULs—much like American professional basketball coaches on scouting tours for the most promising graduates.

The first squadron in a line fighter regiment accommodated the most seasoned pilots and was a dedicated air-to-air unit. The second squadron was made up of less-experienced pilots and performed the regiment's assigned secondary missions. New pilots out of the VVAULs reported to the third squadron, which was in effect a mini–replacement training unit (RTU). Regimental commanders found themselves saddled with as many as twenty-five new pilots at any given moment, facing the long road to mission readiness in an upgrade program that might continue for three years or more. This meant that at least a third of any fighter regiment would be constantly operating in a requalification mode.[14]

This approach to operational conversion entailed a grossly inefficient division of regimental supervisory attention, for it meant that every commander was responsible not only for his unit's combat readiness, but also for the dictates of a rudimentary upgrade program that typically worked at cross-purposes with his mission-readiness needs. This naturally fed a certain cynicism among the more hard-bitten senior pilots, who understandably were more interested in attending to their mission employment concerns than nursemaiding wet-

---

[14]Major General Ya. Yanakov, "From Retrograde to Modern," *Aviatsiia i kosmonavtika*, No. 1, January 1992, pp. 2–3.

behind-the-ears newcomers from the VVAUL system. At a gathering of instructors in a converting MiG-23 unit convened to hear a briefing on the MiG-29's vaunted departure and spin resistance characteristics, one jaded colonel sitting in the back of the room gave vent to such impatience when he muttered skeptically: "Let some nonrated lieutenants fly it. *They* will find a way to spin it."[15]

The pilot upgrade syllabus is embedded in a VVS document called the Combat Training Course (*Kurs boyevoi podgotovki*, or KBP). It features a measured progression through the following rating levels:

Pilot Third Class. This bottom-rung rating typically requires a year of training after VVAUL graduation. It calls for 600 sorties overall, for about 350–400 hours of total time. Upon receiving this rating, the upgrading pilot is cleared to fly day-only training missions with weather minimums of a 750-ft ceiling and a half-mile visibility, and in formations ranging from a four-ship flight to a full squadron of sixteen aircraft. It typically means completing the first 52 syllabus blocks or "exercises" (ranging from one to eight or more sorties) in the day portion of the KBP.

Pilot Second Class. The next rating may take as many as three to four years after graduation. It requires 770 sorties, for around 450–500 hours of total time. Those earning it are cleared to pull both day and night alert duty, are fully instrument flight rules (IFR) qualified, and are authorized to fly supervised ground attack and air combat training missions. They are allotted the same weather minimums as Third-Class pilots, with additional night minimums of three miles visibility and a 1500-ft ceiling. Second-class pilots are cleared for all conventional air-to-ground mission events. Zuyev observed that getting past the written and practical IFR exams and night formation qualification made the Second-Class rating the military pilot's toughest hurdle.

Pilot First Class. This rating requires 1200 sorties, for a minimum of 550 hours of total time. Achieving it often took six to seven years beyond VVAUL graduation.[16] (A USAF pilot reaches this same experi-

---

[15]Zuyev, *Fulcrum*, p. 140.

[16]Although Zuyev made it in only two years and three months, indicating the radical differences in pacing depending on units. Zuyev, *Fulcrum*, p. 154.

ence level within two years or less.) First-Class pilots are fully IFR rated and cleared to perform any published mission event. Weather minimums are 0.9-mile visibility and a 450-ft ceiling. The minimum currency requirement for maintaining a First-Class pilot rating is 50 hours of total flying time a year, including 10 hours of day IFR flight.

Beyond the First-Class rating is a VVS and VPVO designation called "sniper pilot," said to be reserved for a few "highly experienced leaders."[17] It is not a formal aeronautical rating, but rather more akin to a VVS seal of approval for demonstrated professionalism, experience, incident-free flying, and so on. Sniper pilots must have a First-Class rating and a minimum of 1500 hours of total time. They are not necessarily the best pilots in the VVS, but rather those favored by the system or otherwise commanding either special pull or longevity. One VVS general told me recently that the designation is meaningless as far as formally evaluated airmanship is concerned and that the highest proficiency level recognized by the KBP is the First-Class pilot rating.

Finally, at the pinnacle of the VVS pilot hierarchy is yet another rating (or, more correctly, decoration) called "Honored Military Pilot." The few bearers of this coveted honorific typically include the commander in chief, his principal deputies, the commanders in chief of the major commands, and a scattering of fast-burner regimental commanders marked for a rapid climb up the career ladder. It is awarded for special aeronautical and professional accomplishment.

RECENT CHANGES IN THE UPT SYSTEM

One of the most serious weaknesses in the Soviet approach to military pilot selection was its provision that allowed the aspirant himself to pick the aircraft and mission type for which he would compete. This deficiency was spotlighted in a comment by the chief of the VVS's Cadres Training and Assignment Directorate that applicants were expected to indicate to their *voenkomat* the name of the VVAUL they "wished to attend."[18] Such an approach may well have ensured

---

[17]Ibid., p. 112.

[18]Interview with Colonel Leonid N. Pakhnin, "For Those Who Choose to Fly," *Krasnaia zvezda*, February 27, 1993.

that only the most promising and capable cadets accepted into any VVAUL would successfully complete the training program and earn wings. However, because it did nothing to stream the right *kinds* of candidates to those schools where they would be best suited by temperament and abilities to begin with, there was no way of guaranteeing that the best people in the country for fighter training, for example, would be properly identified and vectored toward such training.

The old approach to VVAUL training was sharply criticized in 1990 by Lieutenant General Nikolai Antoshkin, then–air commander for the Moscow Military District. Antoshkin complained that the program was churning out too many graduates who were "neither pilots nor engineers" and whose inadequate airmanship skills were putting an excess burden on line regiments, making the latter, in effect, basic training units to pick up the slack left by a deficient VVAUL syllabus.[19] A deputy VVAUL commander concurred, attributing the poor quality of VVAUL graduates to excessively low required performance levels and "rigid restrictions" on what was allowable in UPT training.[20] This, in his view, meant simply handing off unresolved problems to line regiments. It further constituted a situation that could only be rectified by the top VVS leadership.[21] Evidently the first deputy commander in chief of the VVS, Colonel General Viktor Kot, had no quarrel with that, as attested by his comment that the VVAUL course was "the weakest link" in the VVS's military education system.[22]

A new approach to UPT training was first aired in the summer of 1988 by the VVS's deputy commander for education and training, Colonel General Goryainov. Commenting on the high washout rate of cadets during the primary phase, Goryainov proposed a new look

---

[19]Interview by Major S. Prokopenko, "A New Four-Stage Training Program for VVS Pilots," *Krasnaia zvezda*, October 6, 1990.

[20]Straight-in approaches or long final approaches from an extended downwind leg are standard practice in Russian fighter operations, as was uniformly confirmed by my own flight experiences in the MiG-29, Su-27, MiG-23, and MiG-21.

[21]Colonel V. Sobolev, deputy commander of the Chernigov VVAUL, "Stop Teaching in a 'Retro' Style," *Aviatsiia i kosmonavtika*, No. 3, March 1991, pp. 4–6.

[22]Interview with Colonel General Viktor S. Kot, "We Train the Pilot, We Educate the Person," *Krasnaia zvezda*, March 18, 1993.

at VVAUL selection criteria, with greater emphasis on emotional stability and the ability to refocus attention quickly.[23] Goryainov noted that many cadets who had been deselected shortly after their first solo flight could have been eliminated sooner, more economically, and with less emotional trauma had there been adequate screening criteria to detect their deficiency early on. He admitted frankly that VVAUL candidate selection methods were "weak" and that the VVS needed more support from the human resources community.

By 1988, the VVS had abandoned operational conversion in the VVAULs, whereby fourth-year cadets in the fighter pipeline would complete an initial checkout on the MiG-21 or MiG-23 before being handed off to line units for upgrading and final mission certification. Partly as a cost-saving measure, the schools had already begun instructing on a single aircraft type (either the L-29 or L-39), with graduates proceeding to the equivalent of a USAF replacement training unit (RTU) for initial qualification training on their assigned combat aircraft. During the RTU course, students concurrently earned a Third-Class pilot rating before proceeding to their assigned unit for top-off training.

The purpose of introducing this new RTU, called an Aviation Training Center (*Uchebno-aviatsionniy tsentr*) was to relieve line unit commanders of much of their former conversion training burden. It remains unclear how well the new system has performed in this respect. One weapons instructor in a fighter regiment complained in 1990 that the UPT and RTU phases were still not satisfactorily attending to the full spectrum of training needs, since "a significant portion of the workload for aircrew training" was still being passed on to line units. Consequently, he observed, operational squadrons today "resemble flying school units [and] are stretched to the limit just to upgrade and certify their pilots. Because of this, there can be no advanced training in the intended sense of the term." His implied message seemed to be that the VVAULs and RTUs needed to work on sending a better product to line units so that squadron commanders could pursue an orderly aircrew mission certification plan without

---

[23]Colonel General A. Goryainov, "How to Breed the Next Generation of Fliers," *Krasnaia zvezda*, August 3, 1988.

feeling constantly driven to dedicate extra sorties toward remedial instruction of their new pilots in the basics.[24]

Shortly after the MiG-29 entered service in 1987, the VVS conducted an experiment in which the best graduates from the Chernigov VVAUL were selected to convert directly to the new fighter in the guards regiment at Kubinka.[25] In earlier times, the standard practice was to upgrade only experienced pilots onto new equipment in the initial cadre. The MiG-29, however, proved uniquely forgiving to new VVAUL graduates.[26] Since then, it has been common for most graduates to draw their first line assignment on fourth-generation fighters, since the older types have increasingly been retired from the active inventory.

Initial MiG-29 conversion was provided by the 4th Advanced Tactical Training and Evaluation Center at Lipetsk, the VVS's main fighter weapons and tactics facility about 400 miles south of Moscow. Zuyev participated in the third MiG-29 checkout course, receiving a month of academics on aircraft systems but no flight training. Afterward, he returned to his new unit and was given a local-area checkout, from initial qualification to full mission readiness, entailing 46 sorties.

Undergraduate pilot training in the post-Soviet VVS has undergone a top-to-bottom change. Under the previous system, an aspiring pilot would receive initial flight orientation in a DOSAAF (Soviet paramilitary) aero club during his secondary school years. He would then enter the four-year VVAUL regime, which combined officer, engineering, and flight training through initial conversion to a combat aircraft type. In contrast, the new system approved by General Deinekin entails initial flight orientation in secondary boarding schools, followed by screening and selection for a *five*-year VVAUL program, with flight

---

[24]Major S. Goroshkin, "Disquieting Symptoms," *Aviatsiia i kosmonavtika*, No. 1, January 1990, pp. 10–11.

[25]See A. Gorokhov, "On the Modern Ones," *Krylia rodiny*, No. 8, August 1987, pp. 1–2.

[26]For more on the MiG-29's ease of handling, see Benjamin S. Lambeth, *From Farnborough to Kubinka: An American MiG-29 Experience*, Santa Monica, California, RAND, R-4000-RC, 1991.

training solely on the L-39.[27] Now cadets receive three years of classroom academic instruction. Only afterwards is it determined whether they will continue on to the flying phase, which is now compressed into two years rather than spread out over three as before.

Those cadets cleared to continue in the flying program receive 200 hours of generic training on the L-39. They then transition to their assigned operational aircraft at an RTU. As a result of these changes, according to General Deinekin, those cadets entering the primary flying phase will now be intellectually mature young men, "not 17-year old greenhorns incapable of controlling not only an aircraft but themselves." General Antoshkin added that the VVS has begun "looking toward an RAF-style streaming system. If a fast-jet student fails to make the grade, even at a late stage of training, he has no opportunity of flying any type of aircraft."[28] This represents a major step in the right direction.

The first two boarding schools, collocated at the Yeisk and Barnaul VVAULs, respectively, opened their doors to aspiring high-school students in September 1990. These are not VVS schools, but rather operate as a part of the national educational system. Male students in the 15- to 16-year age bracket who have completed junior high school and are physically qualified for pilot training are eligible for admission, with preference going to those who have received initial flight orientation in DOSAAF aero clubs.[29] Early optimistic portrayals of the boarding school system envisaged 10th-grade students receiving 25 hours of screening instruction in the L-39, with 11th-

---

[27]According to the VVS's deputy chief for education and training, only two of eight planned boarding schools had opened by early 1992, with one reported to be on the verge of forced closure because local authorities had failed to uphold their promise to help finance it. Major General Ya. Yanakov, *Aviatsiia i kosmonavtika*, January 1992, pp. 2–3.

[28]"Moscow's Air Defense Commander," *Air Forces Monthly*, January 1993, p. 21.

[29]Lieutenant Colonel V. Smirnov, "Special Boarding Flight Schools," *Aviatsiia i kosmonavtika*, No. 1, January 1991, p. 18. DOSAAF was disbanded in 1991 and replaced by the Russian Defense Sport-Technical Organization (ROSTO in its Russian acronym). Its chairman noted the difficulty of sustaining flight activities in the face of the funding crisis, but said that its 48 aero clubs nevertheless were training 1500 candidates a year for entry into VVAULs. Major General Aleksei Anokhin, "Our Main Task Is to Educate Patriots," *Military News Bulletin*, Moscow, No. 7, July 1993.

grade students logging an additional 15 hours, plus the initial portions of pre-VVAUL ground school training.

Whatever the underlying rationale, it made little sense from either a training value or cost perspective to offer callow high-school youths such prolonged exposure to a complex aircraft like the L-39 in a mere familiarization program. If this ill-conceived idea ever got past the talking stage in DOSAAF before that organization folded in 1991, it almost surely has not survived the deep budget cuts that have since afflicted Russian military aviation across the board. Even in the best of circumstances, Western aviation professionals would have faulted it as a needlessly expensive and counterproductive extravagance.

The VVS's roster of schools for pilot and other officer specialty training has been trimmed considerably from that of the former Soviet Air Force. Of the twelve original VVAULs, the VVS has been left with only four.[30] Flight schools now include the Kacha VVAUL, named for A. F. Myasnikov, near Volgograd for fighter pilots; the Tambov VVAUL, named for M. M. Raskovoi, for bomber pilots; the Barnaul VVAUL, named for Marshal of Aviation K. A. Vershinin, for ground attack pilots; and the Balashov VVAUL, named for Chief Marshal of Aviation A. A. Novikov, for transport pilots. Navigators are trained at the Chelyabinsk flight school. Military air traffic controllers get their training at the former VVAUL at Yeisk.[31] The Zhukovskii Air Force Engineering Academy in Moscow trains maintenance officers in aircraft and power plant, armament, avionics, and electrical systems. The VVS lost the Chernigov and Kharkov VVAULs and the Lugansk navigator school to other republics in the wake of the USSR's breakup. Yeisk and Orenburg are being reorganized. The Sysansk and Saratov VVAULs for helicopter pilots were transferred to the ground forces when the latter took over all rotary-wing aviation. Two former Soviet VVAULs were combined into a single school at Tambov for prospective multiengine naval and VTA pilots.

---

[30]Similarly, VPVO is now down to two pilot training schools, the Armavir VVAUL, named for Marshal of Aviation P. S. Kutakhov, and the Stavropol VVAUL, named for Marshal of Aviation V. A. Sudets. The latter also trains MiG-31 backseaters.

[31]For a full list, see "Air Force Military Educational Institutions Invite You," *Krasnaia zvezda*, February 27, 1993.

Several years ago, General Goryainov indicated that there was no shortage of VVAUL applicants, that annual vacancies were being filled, and that the schools were operating at full capacity. Yet he conceded a need to raise the level of competition and to improve initial screening in the interest of getting stronger candidates. More recently, there were five applicants reported in 1995 for each vacancy at Barnaul, where prospective Su-24 and Su-25 pilots receive their introductory flight training. Because of the current pilot glut, however, the number of vacancies is almost surely way down. Cadets now receive only 150 hours of primary and basic instruction on the L-39, contrasted with 220–230 hours during the Soviet period and a planned 200 hours when the new post-Soviet VVAUL system was inaugurated.

The fuel shortage and lack of spares have further undermined the intensity and tempo of training. There are 300 students in three training units at Barnaul, but typically no more than twelve of the 80 L-39s at the Kamen-on-Ob auxiliary airfield are mission-capable at any given time. Graduates of the school proceed to the Borisoglebsk RTU for a year of combat crew training on their assigned aircraft type. They remain badly undertrained when they get there, however.[32] At the end of 1992, half of all VVAUL graduates elected to be transferred directly into the inactive reserve. Fifty of the remaining graduates were sent to the Zhukovskii Air Force Engineering Academy for training as maintenance officers.

INSTRUCTOR PILOT SELECTION AND TRAINING

There are three types of instructor pilots (IPs) in the VVS: flight examiners in operational regiments, RTU instructors, and basic flight instructors in the VVAULs. In line units, flight commanders and senior supervisors typically maintain IP status on their assigned aircraft type. Each line regiment offers an IP upgrade course, and IPs in operational units are all First-Class pilots.[33] Those in RTUs all have had previous operational experience. Instructors in the VVAULs occupy the lowest rung on the ladder and are more often than not stuck in a

---

[32]Lieutenant Colonel V. Beltsov, "In the Skies of Altai," *Vestnik vozdushnogo flota*, No. 2, 1995, pp. 44–47.

[33]Conversation with a former Soviet fighter pilot.

dead-end job. A majority of VVAUL instructors are so-called "plowbacks"—VVAUL graduates assigned directly to UPT training duty with no operational experience.

General Goryainov conceded that VVAUL instructors lead a bleak existence, typically working a 10- to 12-hour day and living in cramped quarters offering no privacy to themselves or their families. He also confirmed that IP status in the VVAUL system offers little by way of career prospects. Sorry living conditions at auxiliary airfields drove officers away from any interest in serving as flight instructors. Another disincentive is rotating temporary-duty assignments at remote training airfields, which forces pilots to remain separated from their wives and families often for months on end. Belenko called being a VVAUL instructor "the worst duty conceivable."[34]

The commander of the Kharkov VVAUL offered as his personal view, shortly before the USSR collapsed, that VVAUL instructors should be selected competitively from among Second- or First-Class pilots in line units.[35] General Goryainov, however, confirmed that primary reliance on plowbacks would continue when he announced the inauguration of a new system in which potential VVAUL instructors would be selected from among those graduating cadets found to possess a "special talent" for instruction. These would then be streamed after graduation into specialized instructor training. As for whether cadets with their hearts set on operational flying would ever consent to enter such a program willingly, Goryainov acknowledged that the belief that "those who can't do, teach" remained strong and widespread. He further conceded that only "dire circumstances" would impel a line combat pilot to request assignment as a VVAUL instructor. Volunteers for duty as a first-assignment instructor pilot (FAIP) are extremely rare.

Goryainov disagreed that only operationally experienced pilots would make for good VVAUL instructors.[36] He did, however, main-

---

[34]Barron, *MiG Pilot*, p. 75.

[35]Interview with Colonel V. Shevstsov, "Pilots Who Never Developed: Can We Halt the Attrition from Military Schools?" *Krasnaia zvezda*, March 12, 1991.

[36]In the Israeli Air Force, by way of contrast, it has long been a rule that flight school instructors must have served at least one operational assignment on a combat aircraft.

tain that only First-Class pilots should serve as IPs in the VVAULs. He also rejected the idea that VVAUL IPs picked from line units should remain current in their operational aircraft, since their tactical proficiency could never be kept at the level required by published VVS readiness standards.

Clearly, much post-Soviet adjustment and groping are in evidence here. The VVS seems to be searching for an approach loosely analogous to the RAF's distinction between Qualified Flight Instructors (QFIs) schooled in the full spectrum of basic flight and instrument instruction, and Fighter Combat Instructors (FCIs) limited to a specific aircraft type and focused on mission employment.[37] To qualify as a basic instructor in the VVAUL system now requires a full year of training.

One VVAUL instructor complained that the system that had devolved over decades past had led to a catastrophic decline in the professionalism of Soviet military flight instruction: "The old is forgotten, and anything new that goes beyond the bounds of what is officially allowed is proscribed." IPs are said to search in vain for a legendary manual on instructional techniques written in 1953 by one G. Golubev. Nothing, they maintain, currently exists to replace it. The teaching of instructional techniques has been derided by experienced instructors as comparable "only to the antiliteracy campaign at the beginning of the 1920s." The IP career track itself is portrayed as the ultimate explanation for its lack of professional appeal: "What combat pilot would agree to abandon his line unit for a VVAUL, his combat aircraft for a trainer, his higher pay for low pay, and his

---

Flight school instructors on the Fouga Magister or A-4 maintain mission currency in their operational aircraft.

[37]VVS officers have shown a natural curiosity about the USAF's UPT and continuation training programs. Their image of these programs is often entertainingly distorted. One account wrongly remarked that "only experienced aircrews with substantial flying experience and who have passed special examinations are assigned to strategic aviation." It further commented that physical fitness standards among American youth have declined as a result of "a sedentary lifestyle; over-preoccupation with videotapes, computer games, and riding around in automobiles; excess weight; and drug use." The VVS evidently has a copy of MCM 51-50 for the F-16, since the article reprints ten sections of it. Colonel A. Drozhin and Lieutenant Colonel A. Kokorev, "Aircrew Training in the U.S. Air Force," *Aviatsiia i kosmonavtika*, No. 9, September 1991, pp. 32–35.

agreeable work schedule for the exhausting duty of a training regiment?"[38]

Instructors in the VVAULs handle up to six students at a time, treating their charges much in the manner of combined big brother and watchdog. They not infrequently fly up to nine instructional sorties in a day that begins with a pickup at 0400. The flying is typically unchallenging (Belenko commented from his own experience that after the fortieth loop of the day, "a loop was not so interesting").[39] Like most line units, VVAUL squadrons lack computers to facilitate record-keeping by instructors. Consequently, the squadron commander and his deputy must spend a minimum of three to four hours out of a 14- to 16-hour workday building the next day's training schedule by hand.

The individual instructor pilot remains only a cog in this system, with his every step monitored and "with no initiative permitted in any case." Commanders and other supervisors routinely do the thinking for instructors and second-guess their every suggestion, particularly regarding the elimination of failing cadets. According to Zuyev, considerable pressure was placed on instructors by the school's leadership to shepherd slow learners through the syllabus. VVAUL commanders were reluctant to wash out a marginal performer because of the Communist Party's requirement that VVS pilot production norms be unwaveringly met.

For their part, IPs are "gripped in a vise of documents regulating flight operations" and are allowed little freedom to decide how to teach, what methods to use, or even what to emphasize. Inevitably, this has occasioned a pronounced morale problem. Explained one bitter instructor: "Working with one eye constantly on one's superior officers instills a feeling of fear, never mind any inclination toward showing initiative or creativity. This fear, in turn, engenders a lack of confidence in one's actions. I see this as the reason for the pessimistic and dispirited mood among pilots in many regiments."

---

[38]Lieutenant Colonel N. Gorchakov, "Put In a Word for the Poor Instructor," *Aviatsiia i kosmonavtika*, No. 10, October 1991, pp. 10–11.

[39]Barron, *MiG Pilot*, p. 79.

ISSUES IN TRAINING PHILOSOPHY

With the onset of glasnost, Soviet fighter pilots and test pilots grew increasingly voluble over the self-imposed deficiencies of the VVS's conservative approach to flight training at all levels. A case in point was the discontent that manifested itself over the VVS's summary abandonment of an adequate spin recovery training program. Shortly after the introduction of the L-39 into the VVAUL system, the aircraft showed sluggish spin entry characteristics and, in some conditions of flight, an inability to recover from a spin. Subsequently, the previously mandatory spin training portion of the VVAUL syllabus was eliminated. As of the spring of 1990, spin training in the L-39 remained prohibited, and demonstrations of departure from controlled flight were taken only to heavy pre-stall buffet.

A VVS colonel questioned the sense of this approach, noting that one could teach a student to handle an aircraft to its full maneuver capabilities either by instilling habits that would never produce stall and spin conditions, or by actually demonstrating stalls and spins so that students would truly know their aircraft and what recovery techniques worked best. The VVAULs, he noted, still followed the former approach. Yet that hypercautiousness did not prevent students from flying into unusual attitudes that resulted in departures from controlled flight. This colonel cited a confidential survey of MiG-21 instructors that indicated that many did not fly their aircraft into heavy buffet, even though they might remain well short of the aircraft's stall and departure threshold. Because of this arrested training, he said, they were consistently failing to extract the full extent of the aircraft's maneuver and performance capabilities.[40]

Not long afterward, an experienced test pilot and two Mikoyan Design Bureau engineers concurred. They explained that the ban on spin training in the L-39 was a result both of "subjective views of individuals" and, even more, of the "concepts for air power application

---

[40]He added that pilots with stall and spin familiarization are more likely to teach these techniques to their students, while those lacking such experience emphasize training in departure region avoidance. He said it was time to force an "unambiguous answer" to the question of whether the more aggressive "know your airplane" approach or the more conservative, safety-oriented one was preferable in combat training. Colonel G. Rayevskii, "Should the Spin Be Taught?" *Aviatsiia i kosmonavtika*, No. 3, March 1990, pp. 6–7.

that had taken shape" during the 1960s and 1970s.[41] They further noted that during the years when spin training had been routine, the loss rate of aircraft to departures from controlled flight was low, and that line pilots entered maneuvering engagements with confidence in their ability to handle the aircraft. Spin recovery training in the MiG-15 was standard until the VVS commander in chief banned it, purportedly because the loss rate in the VVAULs was exceeding the rate in continuation training in line MiG-15 squadrons. The trio attributed this negative attitude to the changed approach to VVS and VPVO training which began during the early 1960s, when the main air-to-air mission profile consisted of a single-pass missile attack against a high-speed, high-altitude target that required no hard maneuvering.

With the renewed emphasis on maneuvering air combat in the 1970s, a resurgent premium was placed on the importance of advanced handling skills. Since flying aggressively to angle-of-attack (AOA) limits is intrinsic to maneuvering air combat, it was deemed essential for pilots to have full confidence in their aircraft's departure-region handling characteristics by experiencing them through spin training. Without such confidence, they would never extract the most out of their aircraft's performance, owing to a constant fear of departing from controlled flight.[42] As a first step toward a remedy, the trio recommended establishing a nucleus of instructors in a setting like the civilian test pilot school at the Gromov Flight Research Institute to develop a syllabus for safe spin training for VVAUL cadets.

Another input into this debate noted that full operational leverage could be attained only by flying an airplane to its limits, and that this

---

[41]Hero of the Soviet Union and Honored Test Pilot A. Shcherbakov, with A. Klimov and A. Gorlov, "Teach the Spin," *Aviatsiia i kosmonavtika*, No. 1, January 1991, pp. 10–11.

[42]Retired U.S. Navy Captain Jerry O'Rourke has summarized this thinking in a nutshell in his comment that "a fighter pilot must use his airplane right up to its limits in his routine flying, be it in combat or training for combat. These fine edges between what the plane can do and what it cannot are his ballpark. The mark of the true professional is his ability to get into that ballpark and to drive his enemy out. So he must *use* his airplane and his weaponry right up to these limits. If he doesn't—if he reserves a little cushion for safety, or for the wife and kids, or for any lack of personal confidence—he's not really a fighter pilot, and, when combat comes, he'll soon be beaten by one who is." Quoted in "Fighters That Never Got to the Fight: Part II," *U.S. Naval Institute Proceedings*, April 1982, pp. 76–77.

remained impossible because of "tight restrictions" on such flight parameters as *g* and AOA in the interest of safety.[43] The writer, a retired colonel, noted that VVAUL and RTU students are not systematically exposed to flying their aircraft to the limits. This means that inadvertently approaching the edge of the performance envelope could cause a pilot to aggravate rather than correct an incipient problem. He insisted that mastering "prohibited" flight regimes was dictated not just by the possibility of inadvertently entering them, but because in combat (as Soviet pilots had repeatedly seen in Afghanistan) the instinct for survival will inevitably force pilots into flight regimes beyond published restrictions.

The colonel went on to say that "the existing system for determining restrictions is imperfect and in serious need of correction." He cited the inherent tension between peacetime concerns for safety and wartime concerns for ensuring maximum combat effectiveness, with the former calling for tighter restrictions and the latter for looser ones. He argued that since pilots in combat will surely exceed published limits in the interest of winning—or surviving—they must learn what to expect during peacetime. Revealingly, he noted that after the introduction of the MiG-21 into operational service, there had not been a single case of a pilot recovering successfully from an inadvertent spin. Naturally, because of the great altitude loss involved and the uncertainty of recovery, he did not advocate spin training in high-performance aircraft. He did, however, call for departure-region familiarization in the interest of spin entry recognition and prevention.

---

[43]Colonel (Ret.) Nikolai Lysenko, "How Can Maximum Flight Modes Be Mastered?" *Aviatsiia i kosmonavtika*, No. 7, July 1992, pp. 6–7.

Chapter Five

CONTINUATION TRAINING IN LINE FIGHTER UNITS

As in the case of undergraduate pilot training, what we can say now with respect to day-to-day operations at the unit level is substantial compared with what was known a decade ago, thanks to the revelations of glasnost. By far the most extensive of these have pertained to fighter aviation. Before reviewing them in detail, let us first lay out the dominant beliefs, assumptions, and uncertainties about VVS training and operational style that prevailed among Western observers when the USSR was still a closed system.

By way of background, Western threat assessors first began paying serious attention to Soviet air combat training and force employment during the mid-1970s, roughly coincident with the establishment of the USAF's Aggressor program, realistic large-force exercises like Red Flag and Maple Flag, and the initiation of dissimilar air combat training (DACT) worldwide. For most of the decade that followed, Soviet tactical air activity remained an object of intense interest, as well as extensive guesswork, throughout the fighter community.

Because conclusive evidence bearing on the character and quality of Soviet training and tactics was largely inaccessible, impressions of Soviet air combat prowess and style were based heavily on observed practices of the air forces of Moscow's military client states, notably North Vietnam, Egypt, and Syria. They gained further reinforcement from the informed comments of occasional defectors and from various inferences one could draw from a close reading of articles in the VVS's monthly magazine and other Soviet military writings. From this fragmentary evidence, it appeared that the Soviet pilot was bound by a heavily scripted tactical repertoire that was all but completely dominated by ground control. This, it was generally thought, left him little room for exercising the free-form initiative and adaptability that had long been a hallmark of Western tactical air practice.

There was considerable disagreement, however, over the extent to which that assumed deficiency was an inherent condition of Soviet fighter aviation, let alone a weakness that mattered much in a military establishment that was widely believed to be ready as a matter of practice to trade high loss rates for victory. Intense debates ensued as protagonists for both schools of thought sought to justify their interpretations of the Soviet air challenge. Each school was able to draw sustenance from the inherent ambiguity surrounding much of the limited information on the Soviet air arm and its operational activities.

CONTRASTING WESTERN VIEWS BEFORE THE FALL OF COMMUNISM

Two stereotypes of the Soviet tactical air challenge vied for dominance in the West throughout the later years of the cold war. One was the view typically propounded in official documents like *Soviet Military Power* and formal threat appraisals used to support the Defense Department's annual budget requests. This intentionally conservative view was based largely on such known tangibles as aircraft and weapons capabilities, observed operational performance, force structure and size, and related quantifiables. It maintained that the VVS outnumbered NATO in fighter strength and was making steady advances in its equipment, advances that promised to narrow, if not eradicate, the West's perceived edge in combat prowess.

This worst-case perspective tended to magnify the threat by drawing linear inferences from such proven measurables as range-payload capability and radar and missile operating envelopes and extrapolating from these assumed operating attributes of Soviet equipment, without much thought as to how that equipment might actually be used in a campaign setting. In stipulating that a Su-24 tactical bomber with a standard weapons load could fly, say, 700 nautical miles to a designated target deep inside NATO territory and return home, for example, most analysts in the West failed to ask whether a typical captain in a typical Su-24 squadron commanded the proficiency and adaptability required actually to fly critical portions of that route at low level in marginal weather, in the face of determined NATO opposition, and to find the target and achieve the planned results.

A second view predominated among the ranks of American and allied fighter pilots with appropriate access to give informed consideration to the question. This view held that however impressive the VVS's equipment and order of battle might appear on paper, the Soviet pilot was poorly trained by Western standards and represented a weak link that could be exploited. He was seen as merely a cog in a wheel, a rigid "by-the-book" aviator who deferred to his ground controller on all important decisions, and above all as the product of an operational culture that made a point of teaching him not to do much thinking for himself.

Typical of the latter view was this appraisal by a USAF Aggressor squadron commander well versed in Soviet fighter weapons and tactics: "Exactly how good is the enemy? Is he a ten-foot giant? Not exactly. In fact, without exaggerating, one could place him in the mediocre to poor category when it comes to air combat capability. Certainly his equipment has not improved at nearly the rate ours has. Most important, however, Soviet training is so inferior to ours that this could well be the deciding factor in the outcome of the next conflict."[1]

That appraisal was based in part on the poor Soviet showing a decade earlier, when five Soviet-flown MiG-21s were deftly dispatched by Israeli F-4s and Mirages in an intense five-minute engagement over Suez during the 1970 War of Attrition.[2] It was further supported by the known fact that Soviet air doctrine required most sorties to be flown under the close control of a ground-based mission

---

[1]Lieutenant Colonel Mike Press, "Aggressor Reflections," *USAF Fighter Weapons Review*, Summer 1981, p. 4.

[2]In the words of one of the Israeli F-4 pilots who scored a kill in that encounter, the Soviets showed "a very low skill level" and committed "all imaginable errors" once the engagement was joined. See Colonel Aviam Sela, "A Trap for the Russians," in Merav Halperin and Aharon Lapidot, *Chalifat Lachatz* ("G-Suit"), Tel Aviv, Israeli Ministry of Defense, 1987, pp. 69–72.

supervisor, with the added proviso that if contact with the controller were lost, the mission was to be aborted.[3]

An attempt to discredit this view by a USAF analyst equally well informed on Soviet operational practice took exception to its tendency to dismiss the Soviet fighter pilot as "virtually a puppet, rigidly controlled by GCI and acting as little more than flying artillery."[4] This counterargument maintained that "the situation has changed considerably in the last few years. The Soviets have entered a period of intense study and discussion of their theory of tactics and are beginning to implement changes in their operational training." A comparable claim was reflected in the U.S. Defense Department's 1983 edition of *Soviet Military Power,* which declared that "the Soviets have recently made significant changes in their air combat tactics and training programs. Pilot independence and initiative are now stressed."[5]

Despite heated contention over such questions, the prevailing view held that the typical Soviet fighter pilot was probably up to Western standards in his basic flying ability, yet remained largely untutored in the maneuvering skills and tactical mindset that made the crucial difference between winning and losing in a dynamic, multiparticipant engagement. That image was reinforced by the heavily choreographed and clumsy manner in which Soviet air defense fighters engaged and shot down Korean Air Lines Flight 007 in September 1983. As indicated in detail by the published transcript of air-to-ground communications between the Su-15 pilot and his GCI controller, the intercept was a model of confusion and directed behavior

---

[3]Michael Skinner, *USAFE: A Primer of Modern Air Combat in Europe,* Novato, California, Presidio Press, 1983, p. 122.

[4]Captain Rana J. Pennington, "Closing the Tactics Gap," *Air Force Magazine,* March 1984, p. 83. A year later, Captain Pennington similarly wrote that "rather than painting the Soviet pilot as ten feet tall, we have consistently depicted him as a midget—a dwarf at best. It has long been a matter of reassurance to the Air Force that no matter how many aircraft the Soviets had, the poor skills of their pilots would significantly hamper their ability to use those aircraft effectively." "Another Look at the Soviet Fighter Pilot, *Air Force Magazine,* April 1985, p. 83.

[5]*Soviet Military Power,* Washington, D.C., U.S. Government Printing Office, 1983, p. 43.

from start to finish.[6] It took the pilot a full fourteen minutes to down the 747 after his initial reported visual contact. During this period, he was vectored all over the sky by his controller even as he had the aircraft in sight. At one point during the intercept, he closed to within 6000 ft of the target. But he never positively identified it or showed any other sign of initiative in the situation.[7]

Disagreements persisted almost up to the demise of the USSR over the ultimate meaning of this seemingly rigid operating style for overall Soviet warfighting capability. Most American fighter pilots, however, seemed ready to accept as axiomatic the characterization offered by a U.S. Naval Reserve F-4 pilot: "I have found that asking two U.S. pilots for their tactics in a given situation elicits three different answers. By contrast, it is my understanding that three Russian fighter pilots will all give the same answer."[8]

That impression gained powerful backing from an account in 1986 by a British aviation writer based on interviews with Indian Air Force pilots, who spoke freely of their experiences and frustrations while undergoing MiG-21 conversion training in the USSR during the early 1970s.[9] Although those interviews reflected dated information, they dovetailed nicely with the picture offered by more current indicators, including periodic veiled complaints voiced by line pilots in the VVS's monthly journal. The essence of that picture was that continuity had outweighed change during the intervening years.

According to this report, the VVS followed a syllabus approach throughout the service life of the MiG-21 that assumed the student had an almost complete lack of understanding of fighter aircraft, as well as an entrenched incapacity to learn in any other manner than

---

[6]See "U.S. Intercepts Soviet Fighter Transmissions," *Aviation Week and Space Technology*, September 12, 1983.

[7]Granted, this episode was performed by the separate Soviet Air Defense Forces rather than by VVS fighters, which operated under somewhat looser constraints during vectored intercepts. Nevertheless, it was characteristic of the Soviet military's tendency across the board to deny its combat aircrews much room for exercising individual initiative or independent judgment.

[8]Robert L. Shaw, *Fighter Combat: Tactics and Maneuvering*, Annapolis, Maryland, U.S. Naval Institute Press, 1985, p. x.

[9]Bill Gunston, *Mikoyan MiG-21*, London, Osprey Publishing, Ltd., 1986, pp. 36, 47, 64, 88–92.

through repetitive instruction over a prolonged period. The experienced Indians who had enrolled in this course complained that it progressed at a snail's pace by Western standards and fixated at length on the most inconsequential matters, including proper head and hand movements for various procedures and functions in the cockpit. The schools at Tashkent and Lugovaya also enforced exaggerated operating restrictions, including a ban on instrument flight or entry into clouds, any display of individual initiative, or any other departure from strict and rigidly defined procedures. As an example, the Indians stated that they were obligated to land out of a long, flat approach from a wide traffic pattern and were prohibited from making full-flap landings because of the MiG-21's excessive sink rate. Some instructors prohibited their students from generating roll rates in excess of 90 deg/sec.

This was all most unsettling to the Indians, who were experienced pilots brought up in the manner of the Royal Air Force. Although they did not buck the system, they were facing an imminent war with Pakistan and needed to know the MiG-21 to its limits. Yet they literally had to beg to try anything of tactical relevance or value.[10] In addition to being saddled with excessively conservative rules, they found that their instructors frequently disagreed among themselves over what the rules were. In all, they said, the Soviet approach was to instruct "rather in the way small children learn multiplication tables," with students frequently chanting the correct answers in unison. The report added that "any deep thought about how to get the best out of one's aircraft, or even hack an unusual situation, was simply not part of the syllabus."

Fortunately, the information explosion occasioned by glasnost and the collapse of the Soviet state, as the preceding chapters have shown, has made developments in the armed forces of the former USSR easier to understand in detail. In so doing, it has helped break

[10]The Finnish Air Force, which also acquired and continues to operate the MiG-21, reports a similar experience. According to its former commander, retired Lieutenant General Heikki Nikunen, introductory training on the aircraft for the initial Finnish cadre at Lugovaya included no tactical flying and proceeded so slowly that the Finnish team leader finally called a halt to it and brought the group home to complete the process on its own. The MiG-21 was evaluated at the Finnish Air Force flight test center, and optimum tactics were then developed based on those results. Letter to the author from General Nikunen, April 16, 1993.

down much of the mystery that previously shrouded all but the broadest outlines of Soviet fighter employment practice. The result has been to bring to the surface a number of issues concerning fighter aircrew training, tactics development, and flight operations at the unit level that had long simmered but remained suppressed by the Communist Party's intolerance of dissent. In the process, questions that were hotly debated among Western analysts in the 1980s without resolution have been put to rest by the frank admissions that Soviet—and now Russian—airmen have offered in more recent years.

STANDARDIZATION EVALUATION

VVS fighter regiments operate very differently from their USAF counterparts. Although they are no less overburdened by paperwork and reporting responsibilities, they are far less consistent about honoring published rules in day-to-day practice. VVS headquarters promulgates explicit and detailed training guidelines as a matter of daily routine. Yet their enforcement at the unit level is selective at best and capricious at worst, in either case strongly reflective of the personal inclinations of the regimental commander.

A squadron's normal operating routine is governed by the Flight Operations Manual and the Combat Training Course (*Kurs boyevoi podgotovki*, or KBP). The first is a rough counterpart to the USAF's Multi-Command Manual 11 series, which specifies the number and type of sortie events, ranging from instrument approaches to various air-to-air and air-to-ground weapons delivery profiles, required to be performed by each pilot per training period in order to be considered mission-ready. The second is a structured syllabus progression for new RTU graduates upgrading to full mission-ready status. It consists of over 200 transition, air-to-air, and surface attack syllabus blocks called "exercises," proceeding from initial qualification through two-ship and four-ship to regiment-level events. Each sortie profile specifies carefully defined roles, with no allowance for free play or role reversals in a formation.[11] Regimental surge exercises are

[11]Conversation with a former Soviet fighter pilot.

governed by the Manual of Combat Procedures and Regulations in addition to these two publications.

The main measure of a squadron's effectiveness is its performance in its upgrade program and its maintenance of previously achieved First-Class pilot ratings.[12] Advancement to First-Class status is conducted in accordance with the KBP, which is loosely analogous to the USAF's Graduated Combat Capability (GCC) system. A new KBP for all VVS commands was issued in 1986. An improved replacement was later said by the deputy head of the Combat Training Directorate to be in preparation for release in 1992, suggesting that the syllabus is routinely updated every five years.[13]

In theory, training documents like these are intended to enforce standardization. In reality, regimental commanders set their own standards. As a consequence, a weapons instructor complained that there was no way to ascertain a pilot's ability actually to carry out a maneuver sequence short of taking him up for a check ride that included the maneuver sequence. He implied that there was nothing in VVS fighter aviation practice strictly analogous to the USAF's GCC yardstick that would allow "ensuring a steadily increasing complexity of mission assignments from one training sortie to the next."[14] By 1990, experienced pilots were complaining openly about shortcomings in the application of the training norms specified in the KBP. One lieutenant colonel flatly asserted that VVS fighter readiness existed "on paper only."[15]

While he was commander in chief, Colonel General Shaposhnikov insisted that it was essential to provide a means of systematically tracking the mission events flown by each pilot during a training cycle in order to create a more substantiated basis for averting situations conducive to accidents. Citing a case in point from his own former regiment, he noted how the use of such an approach had re-

---

[12]Ibid.

[13]Letter by Colonel T. Sheshenya to *Aviatsiia i kosmonavtika*, No. 10, October 1990, p. 9.

[14]Major S. Goroshkin, "Disquieting Symptoms," *Aviatsiia i kosmonavtika*, No. 1, January 1990, pp. 10–11.

[15]Lieutenant Colonel A. Zhukov, "Is There a Way Out of the Impasse?" *Aviatsiia i kosmonavtika*, No. 6, June 1990, p. 12.

vealed one pilot to have repeated the same mistake *three times* in a single month. This helped determine where remedial efforts needed to be concentrated to increase that pilot's proficiency.

Day-to-day upgrade training is conducted according to the KBP, whereas a unit's performance during operational readiness inspections is assessed according to published criteria in the Manual of Combat Procedures and Regulations. This inconsistency was said by Lieutenant Colonel Zhukov to be directly responsible for one out of every four mishaps during inspections because of the extraordinary pressure the latter put on regimental commanders to succeed. Zhukov further reported that missions flown in accordance with the KBP did not automatically qualify pilots to carry out missions specified in the Manual of Combat Procedures and Regulations. Although it is hard to say for sure from the limited information provided, this sounds reminiscent of a widely cited problem at the squadron level in USAFE during the late 1980s concerning the challenge of reconciling a squadron's need to comply with Multi-Command Manual 11 (then called MCM 51-50) with its NATO TACEVAL requirements, which likewise often worked at cross-purposes.[16]

Zhukov added that experienced pilots were convinced that the existing system had reached a dead end. Commanders, he said, know implicitly that any deviation from the Flight Operations Manual will result in administrative sanctions. They also understand that they can get by with paying only lip service to their tactical training obligations. Units are barraged incessantly with directives from higher headquarters about the need for improved proficiency. Yet, because of the perverse incentives and conflicting pressures at work, these directives are routinely dismissed as empty declarations. Unit commanders pay little heed to the Manual of Combat Procedures and Regulations in their day-to-day training.

Zhukov's proposed solution was to integrate KBP and mission employment training in such a way that new pilots could be systematically advanced toward mission-ready status in compliance with real tactical needs and a clear concept of operations. He noted how experiments had shown that, given an opportunity to think through

---

[16]For further discussion, see Benjamin S. Lambeth, *Tactical Fighter Training in USAFE: Issues and Options*, unpublished manuscript, April 1990.

their training needs themselves, upgrading pilots would quickly rise above "job-related infantilism" and gain a better appreciation of where they needed remedial work. Such an approach, he said, would end the dangerous situation whereby unit commanders temporarily suspend their normal square-filling (or pursuit of "gross numbers") as the operational readiness inspection (ORI) approaches, only to revert to mindless practices once the stress of the inspection has passed. Zhukov insisted that it was past time to put an end to such "shifting back and forth."

Much of what passed for mission certification training in Soviet fighter regiments was conducted at a fairly rudimentary level in comparison to Western practice. Even in 1988, before the beginning of the VVS's funding crisis, a senior training officer wrote that young upgraders had to be drilled in such simple tasks as visual pilotage, navigation aids orientation, maintaining position in formation, meeting assigned range times, and orderly recovery to a safe landing.

Upon assuming command of the VVS, Colonel General Shaposhnikov proposed a major revamping of its operational practices at the unit level. "Long ago," he said in a 1990 interview, "I became convinced that much, much needs to be reexamined in our system of combat training."[17] For a while, he noted, VVS headquarters had assumed that recurring performance deficiencies uncovered during annual inspections were simply testaments to the faulty leadership of unqualified regimental commanders. Yet those commanders would be reprimanded and replaced time and time again, with little discernible effect on the unrealism that dominated tactical training across the board. Eventually, a conclusion was reached that the system itself was the culprit. In an uncharacteristic attempt to fix the problem, claimed Shaposhnikov, the VVS reached way down into the ranks of the flying community to solicit criticism and suggestions from below, on the premise that "if reform is not to become a child of the apparatus, it must go both up and down."

In particular, Shaposhnikov noted a need to change the VVS's approach to training "so that the pilot can continuously improve himself as a professional." As a first step toward that goal, he proposed

[17]Interview with Colonel General Ye. Shaposhnikov, "The Air Force Today and Tomorrow," *Aviatsiia i kosmonavtika*, No. 8, August 1990, pp. 2–3.

dividing pilots into proficiency categories such as "A, B, C, and D," so that "when one makes it into the last group, it will mean in fact, and not just on paper, that he knows everything. . . . As it is, we all rush hastily to get pilots to the First-Class rating, but we don't achieve a stable level of skill."

A related innovation proposed by Shaposhnikov was a more rational apportionment of flying hours among pilots, so that those on the low end of the annual allocation could maintain at least enough proficiency to remain safe, if not fully mission-ready. He cited as an example a notional First-Class pilot on regimental headquarters assignment who flew only 70 hours a year for several years in a row and then somehow surprised everyone when he went to the range and proved himself incapable of putting bombs on the target. He further noted that such pilots caused 60 percent of the VVS's mishaps.

Shaposhnikov implied that he had something in mind roughly comparable to the USAF's GCC approach to proficiency rating when he spoke of a need to "change the whole system." The challenge, he said, was to "define the readiness categories that pilots will have to go through" and then to determine, "legally and methodically," permitted independence levels for pilots in each category. Shaposhnikov seemed to be visualizing a system involving graduated levels of certified aircrew capability and at least some mission specialization within units. Under existing rules, he said, a pilot would typically fly to the range a few times and be declared competent to attack any ground target. That, he pointed out, was a major part of the problem, in that it tended to overlook the different skills needed to execute progressively more demanding mission tasks. By way of a better approach, he suggested that "if, in the future, we will have in a squadron, for example, a flight of experts at disabling airfields, a flight for destroying Hawk antiaircraft missiles, and so on, then we can characterize its combat readiness without bias."

To illustrate how such a classification scheme might work in practice, Shaposhnikov noted that area targets could be lumped together in group A, with smaller or harder-to-hit objectives assigned to progressively higher categories. Then, were a pilot to be working at the performance level associated with, say, group D, he would truly and demonstrably be "capable of a lot." In all events, said Shaposhnikov, the goal must be "to get away from the practice of averaging everything out in our training."

OPERATIONAL STYLE

VVS fighter operations contrast markedly with the familiar training patterns of the USAF and most other Western air forces. The biggest differences are an unbending top-down imposition of inhibiting rules for almost every circumstance of flight, a continued denial of much latitude for individual initiative on the part of pilots, and a dominant role played by the ground command post (*komandniy punkt,* or KP) and the flight director (*rukovoditel' poletov,* or RP) in overseeing both upgrade and continuation training.[18]

In a striking testament to its rigid and stereotyped approach to mission preparation, the VVS uses a practice called "dismounted training," whereby pilots take to the flight line and, with model airplanes in hand, literally walk through each maneuver and event in a planned training mission. Such practice has been observed on Soviet television with pilots in full flight regalia actually lining up in echelon at the end of the "runway"; walking forward to mimic the takeoff roll; calling for afterburner; and proceeding to climbout, departure to the work area, and execution of each mission event in preplanned sequence. However useful this may be in helping pilots visualize what they are about to perform in the air, it reflects a training philosophy based on scripted mission planning and rote memorization that is fundamentally alien to Western practice.

Although the VVS began receiving its initial batches of fourth-generation MiG-29 and Su-27 fighters during the late 1980s, Colonel General Shaposhnikov confirmed that its operational style continued to rely heavily on direction from the ground. He admitted that the new equipment had enabled significant "adjustments in the training process and in flight operations control group (GRP) practices."

---

[18]That autocratic style of command and its associated diminution of the role of the individual goes at least back to World War II days. For some fascinating insights into its early manifestations as far down as the flight-leader level, see Generalleutnant Walter Schwabedissen, *The Russian Air Force in the Eyes of German Commanders,* USAF Historical Studies No. 175, USAF Historical Division, Air University, Maxwell AFB, Alabama, June 1960.

However, he went on to say, "the man at the console continues to play a deciding role." That statement spoke volumes about how far the VVS had yet to go before it could claim to have matched the independence that has long been routinely granted to Western fighter pilots.

GRP is an abbreviation for *Gruppa rukovodstva polyotov* (literally, "flight supervision group"). This is the approximate VVS analogue of a USAF fighter wing command post, but its role is far more intrusive and overbearing. It features a glassed-in control center with a clear view of the ramp, taxiway, and runway at each base. It is manned during flight operations by a flight director (generally a senior pilot with the rank of major or higher), a radar controller, and a regimental-level supervisor. It combines the USAF functions of supervisor of flying (SOF), GCI, control tower, and runway supervisory officer (or mobile control). Most important, it makes many decisions for the airborne pilot that are typically made in the cockpit in Western practice.

Part of the reason for the arrested development of VVS tactical training is this heavy reliance on the ground tactical control officer, whose responsibilities range from seeing to the successful completion of an intercept to making the key decision calls during in-flight emergencies. The ground controller also plays the dominant role when it comes to such elementary matters as deciding when to select afterburner during an intercept. In what would have been a gross affront to the professionalism of Western fighter pilots, a junior VVS medical officer, of all people, once commented how Soviet pilots would typically get themselves into a tight corner, only to discover, to their relief, that in such "critical moments, the controller's calm, even voice and precise instructions [were there to] help pilots maintain composure and successfully extricate themselves from their situation."[19]

By one informed Western account in mid-1991, free-form air combat training remained the exception to the rule in Soviet fighter squadrons and generally consisted of visual setups in cases when GCI was overtasked. MiG-29 tactics differed little from MiG-23 tac-

---

[19]Senior Lieutenant of Medical Services O. Rybnikov, "What Was Said Over the Radio," *Aviatsiia i kosmonavtika*, No. 10, October 1989, pp. 8–9.

tics, and units rarely practiced GCI intercepts against targets below 2000 ft.[20] Command post controllers closely monitored and managed each training sortie. When one pilot en route to a weapons range wandered off his assigned heading only six minutes into the flight, the tactical control officer radioed him: "83, you're drifting right 25 degrees off course. Return to your proper heading!"[21] Typically, if the mission goes well, the equipment is applauded. If something untoward occurs, the pilot receives the blame.

The underdeveloped ability of VVS fighter pilots to respond quickly to changing conditions was borne out indirectly by the exposure of USAFE aircrews more recently to former East German MiG-29 pilots, who by one account "are told where to fly, when to turn on their radar, when to shoot, and when to come home."[22] It has been confirmed by the Luftwaffe's commander in chief, Lieutenant General Hans-Jorg Kuebart, who noted that MiG-29 pilots retained by the Luftwaffe from the East German Air Force following unification had to be requalified because of their differing tactical methods.[23]

Even when he was but a lowly regimental commander in 1975, then–Lieutenant Colonel Shaposhnikov cited several accidents that had occurred because of an inability of pilots to adapt to a change in plans after the mission briefing.[24] By the time he had moved up to become air commander for the Group of Soviet Forces in Germany (GSFG) thirteen years later, Shaposhnikov noted how time-honored practices were hard to abandon, notwithstanding the new license apparently conferred by perestroika: "Born of the time of stagnation, such obstacles as oversimplification, formalism, overcautiousness, and fear of accepting responsibility for a demanding task remain al-

---

[20]Conversation with Air Vice Marshal R. A. Mason, RAF (Ret.), citing "authoritative RAF sources."

[21]Colonel V. Dudin, "Both Strength and Precision," *Aviatsiia i kosmonavtika*, No. 10, October 1991, pp. 6–7.

[22]Quoted in Michael Smith, "East Meets West as MiGs 'Fall' to Soesterberg F-15s," *Air Force Times*, October 14, 1991, p. 25.

[23]Quoted in Pavel Felgengauer, "We Fly Only on Aircraft: The Air Force Commander in Chief Meets the Luftwaffe Commander in Chief," *Nezavisimaia gazeta*, July 16, 1992.

[24]Lieutenant Colonel Ye. Shaposhnikov, "Responsible for the Flight," *Krasnaia zvezda*, September 10, 1975.

most insurmountable in the path toward accelerated progress in combat training."[25]

Shaposhnikov acknowledged the stultification and lack of initiative that largely accounted for the VVS's shortcomings in adaptability, and he stressed the need to get rid of those shackles on VVS effectiveness. As a case in point, he cited the decline in one unit's readiness level because of its leadership's mindless pursuit of "statistical indicators" of performance. That led it to oversimplify its mission tasking and eliminate the most challenging training events, all "masked by the excuse of flight safety." All too often, he complained, "slogans and calls for restructuring" were used in place of what was truly needed to shake the VVS out of its complacency and bad habits, namely, "professional competence, purposefulness, activeness, responsibility, imagination, and a commander's boldness in making decisions."

More recently, two Ukrainian Air Force officers confirmed what they called the "conservatism" of flight training in the former Soviet Air Force, indicating that "greater priority was placed on flight safety than on mission effectiveness." As a result, they said, the typical line pilot only managed to master about half the operational capability of his aircraft. These former Soviet pilots charged that training in the Russian VVS continued to proceed "not like it does abroad, but rather by methods from a half-century ago," and they confirmed the stories of pilots "actually walking through an upcoming flight using a hand-held model airplane." They also acknowledged that "Russia has come to understand the erroneousness of this approach."[26]

SCHEDULING AND SORTIE APPORTIONMENT

The types of sorties flown in day-to-day VVS training are determined by the regiment's annual training plan. In preparing a day's flying schedule, the deputy squadron commander assigns each pilot appropriate mission profiles drawn from the KBP. Weather will fre-

---

[25]Lieutenant General Ye. Shaposhnikov, "Devalued by Time," *Aviatsiia i kosmonavtika*, No. 3, March 1988, pp. 4–5.

[26]Lieutenant General (Res.) Nikolai Kryukov and Colonel Nikolai Litvinchuk, "What Is on the Horizon for the Ukrainian Air Force?" *Narodnaia armiia* (Kiev), July 3, 1993.

quently dictate what complexity of sorties can be flown on any given day. As in Western air forces, the VVS maintains pilot weather minimums for landing. These vary from pilot to pilot depending on their aeronautical rating, often with further restrictions assigned by regimental instructors based on assessed instrument proficiency as determined through simulator and flight evaluations.[27]

Colonel General Shaposhnikov noted that event scheduling and sortie apportionment needed to be better informed ahead of time by such considerations as available flying hours, allowable fuel and training munitions, weapons range availability, and annual weather patterns to ensure "a constant and stable tempo of training and the avoidance of crash programs." For determining which pilots were qualified to fly which sortie profiles, he said that what was needed was not meaningless measures of aggregate activity, but rather hard data on the "actual readiness of each pilot and navigator from each unit, from the two-ship element all the way up to the full squadron."

Shaposhnikov more than once stressed the need for careful planning of a unit's training so as to eliminate problems otherwise created by frantic "last-minute and catch-up rush work." This implied that the VVS has standard procedures for the scheduling of training events, but that individual regimental commanders frequently fail to make proper use of them. "It is essential," said Shaposhnikov, "to make sure that each and every VVS officer knows what he will be doing today, tomorrow, next week, and next month." Furthermore, he added, "once a plan or schedule has been agreed to, it should not be carved up on a daily basis just because of considerations driven by the multitude of things that need attending to that day." This concern is not unfamiliar to USAF wing and squadron commanders, who likewise feel periodic tension between the need to stick to an established training plan and recurrent pressures to modify the flying schedule at the eleventh hour so as to accommodate this or that need of the moment, whether it be a function of weather, maintenance nondelivery, or other contingent circumstances.

Shaposhnikov deplored the tendency of supervisors to schedule pilots for mission events for which they were not qualified. He cited

[27]See Colonel V. Skrynnik, "Each Has His Own Minimum," *Aviatsiia i kosmonavtika*, No. 10, October 1991, pp. 28–29.

one case in which a First-Class pilot was cleared to fly a training sortie in a single-seat fighter, even though he was not current for the scheduled events on the mission. The authorizing squadron commander had assumed that because of his seniority and experience, this pilot "surely must be able to carry out the task." The commander also figured that it would waste a valuable training asset to fly the pilot with an instructor in a two-seater that was badly needed elsewhere on the schedule for upgrading younger pilots. Indeed, said Shaposhnikov, the pilot himself knew that his upcoming flight was in violation of regulations. Yet he pressed ahead anyway out of misplaced self-confidence and flew a sortie for which he was unprepared, setting himself and others up for a potential mishap.

Shaposhnikov also faulted the short-sighted planning that allowed pilots to lose mission currency to begin with and the way in which some units shamelessly justify last-minute changes to previously settled flight schedules using the lame rationale of "reestablishing currency." Although that was an honest explanation, it begged a more important question in Shaposhnikov's view. "Who allowed you commanders," he would ask, "to organize your combat training so as to let your pilots lose their skills in the first place?" There was no more sure-fire way of flirting with accidents, he observed, than to be in a daily mode of constantly having to rearrange sorties and mission events in compensation for poor foresight.

Shortly before the final withdrawal of Soviet forces from Germany, the air commander for the Western Group of Forces (WGF), Lieutenant General Tarasenko, reported that German flight restrictions had kept his pilots from night flying for four months in 1991, and they were to be similarly restricted for six months (from April 15 to October 15) in 1992.[28] Scheduled day missions were directed to recover before 1800 local time, almost two hours before official sunset.[29] Because of these restrictions, WGF sorties in 1992 were apportioned at the regimental commander's discretion, such that the

---

[28]Interview with Lieutenant General A. Tarasenko, "In a Holding Pattern: Comments on Problems of the 16th Air Army," *Krasnaia zvezda*, March 27, 1992.

[29]General Tarasenko also said that his pilots had been directed by the German government to remain above 2000 ft during low-level training flights, complaining (incorrectly) that USAFE aircrews in Germany were still allowed to operate entire four-ship flights down to 250 ft.

youngest pilots and those mission-ready aircrews representing the VVS's core talent pool were given the greatest amount of time. Other Band-Aid fixes included reducing the average duration of scheduled sorties, eliminating repeat passes at the range, combining multiple mission events on a single sortie, curtailing afterburner use and flight into marginal weather to save fuel, and greater reliance on simulators, even though Russia's defense industry had radically curtailed their manufacture and technical support.

In a nutshell, said General Tarasenko, WGF training had been reduced to mission scheduling "according to available fuel rather than operational need." Flight restrictions imposed by the German government were partly offset by periodic unit deployments back to Russia, with week-long detachments to Russian tactical ranges for live weapons training. Those deployments accounted for 37 percent of the WGF's total flight time and 50 percent of its night-flight time in 1992.

During the early 1980s, Soviet fighter pilots flew an average of 130–140 hours a year, with Second- and Third-Class pilots flying more than First-Class pilots because of their greater need to build time and aeronautical maturity. Since 1989, this trend has been reversed, with experienced pilots flying the greater number of hours so that the VVS can keep its dwindling total force as combat ready as possible. Today, because of the near-collapse of operations and support funding for all services, continuation training has ground to a halt in many VVS units. Line pilots in those regiments that still maintain any contingency tasking to speak of are now hard-pressed just to retain their basic landing and instrument currency.

A TYPICAL FLYING DAY

In normal circumstances, the daily flow pattern at a Russian fighter base differs considerably from that of a USAF or NATO base.[30] Regiments operate in two shifts six days a week, with assigned squadrons typically alternating between fly days and maintenance down days. For a squadron working the second shift, flying days are

---

[30]John Barron, *MiG Pilot: The Final Escape of Lieutenant Belenko*, New York, McGraw-Hill Book Company, 1980, p. 1.

Tuesday, Thursday, and Saturday, with maintenance and next-day mission planning conducted on Monday, Wednesday, and Friday.

The week's first maintenance and planning day begins Monday morning at 0900, when the squadron commander, his deputy, and his flight commanders meet to review the previous week's events and plan the coming week's flying program. Much of the latter is drafted the Sunday before by the deputy squadron commander, one of several regimental supervisors who, by virtue of their responsibilities, bear the honor of working a seven-day week. The squadron operations officer then prepares the next day's flying schedule, assigns sortie events for each pilot, and arranges for needed support services, such as en route air traffic control, navigation aids, and range control officers.[31]

Tactical control officers review the flight schedule and determine which personnel will be needed in the command post during flight operations. They further attend to such matters as determining takeoff and landing times and arranging for airspace deconfliction. They keep detailed notebooks listing preplanned maneuver sequences to be flown, as well as airspeed restrictions, recovery fuel levels for each scheduled sortie, fuel consumption rates at different altitudes and for different mission profiles, alternate airfields, pertinent notices to airmen (NOTAMs), and so on.

Pilots meet with their flight commanders from 1000 to 1100 to review flight recorder tapes from the previous week. At 1100–1130 the squadron commander announces mission areas to be emphasized, and the deputy commander assigns each pilot his next day's flight profiles from the KBP. From 1130 to 1300, pilots use these to develop detailed and highly scripted mission plans. The latter are handwritten in notebooks, to the point of actually diagramming the flight path and maneuver sequence to be flown, with every turn and airspeed at each critical point in the mission accounted for in advance. Depending on the sortie's complexity, these write-ups will be approved by either the squadron commander or his deputy. Pilots are then asked to calculate their expected fuel consumption specifics for each planned sortie, often under tight time constraints, submitting

---

[31]Guards Major V. Bazhenov, "Navigation Support of Flight Operations," *Aviatsiia i kosmonavtika*, No. 10, October 1988, pp. 20–21.

their results upon completion to be checked and approved by the tactical control officer.[32]

Lunch is taken from 1300 to 1500, after which mission planning resumes. At 1600, pilots go through a cockpit familiarization review. From 1630 to 1700, they then walk through their planned missions from takeoff to landing. Fighter units also conduct regular classroom instruction in navigation and IFR procedures, including such basics as how to read the instrument landing system (ILS) indicator on approaches. Each squadron has a "methods council" which vets and approves suggestions from instructors and other supervisors and decides on training approaches and areas to be emphasized. The planning day ends with an all-officers' meeting at 1700 with the regimental commander.[33]

On fly days, the squadron commander and flight commanders meet at 0540 to review the day's planned events. Aircraft scheduled to fly are towed from their shelters to the flight line by trucks, which also serve as ground power units for the aircraft. This is followed by a 15- to 20-minute regimental weather reconnaissance mission in a two-seater to check out conditions at the ranges and operating areas scheduled for use that day.

Aircrews arrive at the squadron by bus from their apartments between 0600 and 0700 and take breakfast at leisure in the officers' mess. This is followed by a mass briefing of all aircrews from 0730–0800, during which the regimental commander announces his expectations and assigns broad tasking and safety reminders. Pilots then meet with the meteorologist and tactical control officer for a full weather briefing and overview of objectives for the day's planned sorties.[34]

This is followed by a ground-controlled intercept (GCI) briefing and squadron-wide review of safety and other rules of engagement. This

---

[32]One instructor voiced contempt for this ever-present "Notebook for Immediate Flight Preparations," with its detailed schematics of maneuvers, tiresome explanations of the proper way of executing them, safety rules, techniques for correcting deviations, and so on. He groused that such "orthography lessons" were oppressive enough to deny even a pilot of his experience any freedom of action during a sortie.

[33]Conversation with a former Soviet fighter pilot.

[34]Zuyev, *Fulcrum*, p. 124.

is the only time that VVS pilots have any face-to-face contact with their radar controllers. With respect to the actual conduct of intercept missions, GCI can offer suggestions, but flight commanders have the last word. Most controllers are not trusted to give accurate target information and vectors. Contrary to widespread assumptions in the West, basic intercepts and maneuvering engagements using self-setups are frequently practiced without GCI support.[35]

Every pilot then musters to the regimental physician in a hangar dispensary for a routine medical check before being cleared to fly. Pilots are examined for blood pressure, temperature, pulse, respiration, and general suitability to fly. This is mainly a check for alcohol abuse, which at times has been severe in the VVS, particularly in remote areas.[36] Flight surgeons will normally turn a blind eye if aftereffects of alcohol are detected, declaring the affected pilot down for duty not including flying (DNIF) that day for innocuous reasons such as nasal congestion or an ear infection. However, if outright intoxication is suspected, this means automatic grounding and possibly severe disciplinary action.[37]

Following medical release, pilots scheduled for the first launch go through a final cockpit and emergency procedures review, after which they proceed to life support to draw their helmets and a personal-issue 9mm semiautomatic pistol to carry in the cockpit. This is followed by a walk-around inspection of the aircraft, a brief exchange with the aircraft's maintenance officer, and start, check-in, and taxi for a ritual 0900 takeoff. During taxi to the runway, fighters stop at a

---

[35]Conversation with a former Soviet fighter pilot.

[36]Soviet aircraft cooling and braking systems use pure grain alcohol, which is readily available to pilots from commanders on down. Pilots call it "white gold." The MiG-25 carried 1000 lb of it, leading pilots to refer to the aircraft jokingly as "the flying restaurant." Barron, *MiG Pilot*, pp. 81, 97.

[37]Zuyev insisted at one point in his memoirs that alcohol abuse was rare among Soviet fighter pilots. He also expressed surprise to learn that American and RAF pilots "almost ritually frequented their officers' club bars every evening after flying" and evidently considered drinking to be "a sign of masculinity." Yet elsewhere, he spoke of a lethal indulgence among Soviet pilots called "polar bear," in which the pilots would sit around a table with tumblers of alcohol in front of them, bet money on each shot, and wait for someone to shout "polar bear." The pilots would then down the shots and dive under the table to hide from the imaginary bear. This would typically continue for hours, with the winner being the sole survivor who could pick his way out from underneath the table and walk away. *Fulcrum*, p. 123, 175.

last-chance maintenance check point, just as is routinely done in the USAF.

When fuel allocations were not a constraint, pilots typically flew three or four sorties each flying day.[38] Alternately, they might fly six simple sorties or three more complex profiles. Either way, this meant that pilots and maintenance personnel on flying days were constantly on the go, in a frantic tempo of operations that a former Soviet pilot described to me as *sumashedshiy* ("crazy").

Because most Soviet fighters are fuel-limited, sorties last generally no more than 30 minutes from takeoff to landing.[39] Touch-and-go landings are infrequent. Maneuvering zones and weapons ranges are usually not far from the base, making for little wasted out-and-back time. An oval-shaped air-to-air zone is normally situated within 15 to 20 nautical miles of the field.[40] VVS squadrons typically fly formed flights with the same pilots in each flight. Paired wingmen are the rule and pilot substitutions are rare.

The first sortie is normally down by 0940 and is followed by a second breakfast and short mission debrief with the flight lead and GCI. Scheduled takeoff for the second go is 1020, with a third takeoff at 1140. This first flying shift ends at 1300 with a mass debriefing with the regimental training officer, followed by initial preparations for the next day's mission planning. A second flying shift then takes place from 1300 to 1900. Any pilot who demonstrates identified problems in need of attention on any of his sorties that day may then fly again with an IP in a two-seater for remedial work to correct them.

Typically there are 24 aircraft on a regiment's daily flying schedule. Single-seaters fly five or six turns a day, with two-seaters flying more because of the greater demand on them. A regiment might fly up to 180 short sorties in the course of a single routine training day. IPs fly

---

[38]Ibid., p. 14.

[39]The MiG-29, for example, carries less than 7000 lb of internal fuel, as opposed to 11,000 for the F-15. In my MiG-21 and MiG-23 flights at Ramenskoye, we flew in each case in clean configuration with full internal fuel. With normal afterburner use for an advanced handling demonstration no more than 25 nautical miles south of the field, we landed in each case with minimum fuel within a half-hour. (The Su-27, of course, is a conspicuous exception to this rule, with internal fuel capacity of 22,000 lb.)

[40]Zuyev, *Fulcrum*, p. 23.

more because they are constantly upgrading newcomers. Each squadron maintains one spare for every three aircraft on the flying schedule.[41]

During peak periods, aircraft are turned quickly. Six sorties a day per scheduled aircraft is not uncommon. Two-seaters are flown especially hard, typically with a different upgrading pilot in the front cockpit on each successive sortie.[42] To conserve fuel, commanders are encouraged to take every measure possible to minimize the hold time for aircraft awaiting takeoff and landing clearance. The fuel shortage has further occasioned an increase in the scheduling of multi-event sorties and, whenever practicable, the shutting down of engines and towing of aircraft back to parking after landing.[43]

VVS fighter bases use tactical call signs. Radio frequencies are all preset and are limited to discrete air-to-air and selected ground channels, such as the regimental command post, a GCI handler, and departure and destination controllers for cross-country flights. Pilots are assigned a different three-number personal call sign each quarter. This is prefixed by a two-digit regimental number, which is normally never used.[44] Air-to-air missions may have opposing fighters talking to separate GCI controllers sitting side by side at a common radar console but assigned individual frequencies.[45]

Pilots are taught to maintain radio discipline by keeping voice communications to a minimum. The element or flight lead's signal to the wingman for engine runup prior to brake release is full aft stabilator deflection. Formation takeoffs maintain loose spacing within elements, with the wingman deployed on the runway 45 ft to the right or

---

[41]Conversation with a former Soviet fighter pilot.

[42]Rene Van Woezik and Tieme Festner, "Bear Tracks in Germany: The Soviet Air Force in the Former German Democratic Republic," *Air International*, October 1992, p. 210. A former Soviet pilot told me that a maintenance supervisor who truly wanted to discipline one of his errant subordinates would assign him to a two-seater because of its higher utilization rate.

[43]Colonel Yu. Kuzmin, "Worth Its Weight in Gold," *Aviatsiia i kosmonavtika*, No. 12, December 1992, p. 10.

[44]Zuyev, *Fulcrum*, p. 23. When I flew the MiG-29 at Kubinka, our call sign was 817. Later in the MiG-23 at Ramenskoye, it was 550.

[45]Ibid.

left of his leader, depending on wind, and staggered 75 ft aft, making almost for paired single-ship takeoffs.

The runway environment itself is rough and consists of large, precast concrete blocks laid side by side with tar joints. Debris conducive to engine foreign-object damage (FOD) if sucked down an inlet is often present and apparently uncontrolled. Zuyev reported that during rainy weather, the concrete blocks would shift position and swamp water would come squirting up through the joint lines. This would, of course, be unacceptable in Western air forces.

Zuyev reported one Frontal Aviation tactical practice even worse than the old USAF welded-wing formation, with the wingman wired 150 ft or less off his leader entering a fight. He also, however, alluded to resolution-cell tactics in noting that two MiG-29s might approach the merge in such close proximity that they would appear to the opposing GCI or air intercept radar as a single target. He further noted the use of odd and even altitudes by opposing flights for airspace deconfliction to minimize the chance of a midair collision during head-on engagements.[46] And he indicated a minimum altitude (or "hard deck") of 1800 ft for maneuvering during intercept training.

Some reported VVS operating practices would be considered by the USAF to be both tactically unsound and unsafe. In air-to-air setups, for example, MiG-29 pilots often enter the fight with the master arm switch on and the missile launch button depressed so that the captive missile will automatically register a simulated firing once the infrared search and track sensor (IRST) or helmet-mounted sight locks up a target within lethal parameters.[47] Relatedly, Zuyev reported that he always flew MiG-29 air-to-air training sorties with a full load of 30mm cannon rounds.[48] Either practice would almost surely be deemed by any USAF commander to be an invitation to eventual disaster.

On the other hand, the impressive maneuvers performed by Russian test pilots on the international air show circuit since the MiG-29's

---

[46]Zuyev, *Fulcrum*, pp. 24–26.

[47]Ibid., p. 22. This may explain how one Iraqi MiG-29 apparently shot down his own wingman during Operation Desert Storm.

[48]Ibid., p. 19.

first Western appearance at Farnborough in 1988, notably the tail slide and "cobra," remain prohibited for line VVS pilots because of the danger of putting their airplanes into departure-prone situations. On this point, an article by a senior test pilot and two Mikoyan engineers described the controlled nature of the dramatic cobra maneuver and its susceptibility to producing a departure from controlled flight if not properly executed. They also dispatched the common misconception that the cobra constituted an effective last-ditch guns-defense maneuver, rightly noting that any fighter executing such a maneuver at only 200–250 KCAS (the proper entry airspeed) would "reduce its speed in relation to an attacker insignificantly, while sharply increasing the area to aim at"—in effect rendering itself an airborne strafe panel. They added that the maneuver would also stagnate the aircraft's energy to a point where it would "lose any subsequent opportunity to perform even limited maneuvers."[49]

The collapse of funding for fuel and other consumables that now has the VVS on its knees dates as far back as the late 1980s, when the faltering Soviet economy first began to affect the training regime of line VVS units.[50] Even in 1989, squadrons were reportedly standing down for two-week intervals at a time for lack of fuel, with some aircrews

---

[49]A. Shcherbakov, A. Klimov, and A. Gorlov, "On the Road to Supermaneuverability," *Aviatsiia i kosmonavtika*, No. 9, September 1991, pp. 12–13. The "cobra "maneuver, first popularized by Sukhoi test pilot Viktor Pugachev at the 1989 Paris Air Show and since performed routinely at air shows by both the Su-27 and the MiG-29, is mainly a demonstration of aircraft stability and aerodynamic efficiency at high angles of attack. The maneuver involves a full-aft stick snatch for two to three seconds, followed by a release of back pressure and full forward stick to initiate the recovery. It is executed from level flight at 250 kts or less. An AOA of 90 degrees or more is generated during the maneuver, providing enough instantaneous energy bleedoff to force a close-in attacker, at least in theory, into an overshoot. To achieve these parameters, the pitch limiter must be overridden by the pilot. The maneuver has been extolled by some Russian publicists not just as a last-ditch guns-defense technique, but also as a means for executing a snap-up missile attack (see, for example, Colonel A. Andryushkov, "The Pugachev Cobra," *Krasnaia zvezda*, June 13, 1989). Mikoyan's former chief test pilot, Valery Menitskii, has dismissed it as a "circus event."

[50]Even during the Soviet military's best days, fighter pilots did not routinely fly as many hours a year as their Western counterparts. For example, Lieutenant Colonel Timur Apakidze, one of the first Russian Navy pilots to carrier-qualify in the Su-27K aboard the *Admiral Kuznetsov*, took twenty-three years of service to accumulate 2500 hours total time. See Captain Second Rank V. Pasyakin, "Salamander Clears a Landing: Naval Pilots Have Landed on the Deck of the Carrier *Kuznetzov* for the First Time," *Krasnaia zvezda*, October 19, 1991.

flying only two or three times during an entire month.[51] In some instances, such interruptions were attributed not to a lack of fuel per se, but rather to a unit's mismanagement of its fuel allocation.

In the Soviet Air Force, jet fuel was apportioned quarterly against a regiment's projected flying hours. Often weather or other unforeseen constraints, such as lapses in aircrew currency, would cause units to fly fewer than their planned hours during a given training cycle. If aircrews needed to fly more during the subsequent cycle to make up for lost training, commanders would often feel bound by a literal interpretation of rules against using fuel over and beyond that quarter's allocation. The problem was not a shortage of fuel but poor resource management due to blind and unimaginative adherence to standard procedures.

Only around 65 percent of the flying in VPVO fighter units is dedicated to mission-related events. The remainder is taken up by such ancillary activities as upgrading of new pilots, weather reconnaissance, and providing targets for other interceptors. A typical MiG-31 intercept sortie lasts two hours, with the intercept portion itself taking slightly less than an hour. The remaining flying time simply produces a waste of fuel. As an economy measure, the VVS and VPVO have both taken lately to using older fighters like the MiG-23UB, or even the L-39, for operational support flying not directly related to mission readiness. A single fuel load for a MiG-31 can generate 10–12 sorties of equal duration in an L-39.

In both the VVS and VPVO, many upgrading pilots are now stuck in a situation in which they cannot qualify for a First-Class rating even in six years because of the insufficiency of fuel.[52] Worse yet, it has become hard even to retain one's *existing* rating given the low number of available flying hours. As a result, many of Russia's most experienced pilots are now facing the near-certainty of losing their currency certifications and being forced to requalify. Lamented one

---

[51]Senior Lieutenant V. Shurygin, "A Reader Poses a Question: Excess Fuel," *Krasnaia zvezda*, September 13, 1989.

[52]Under normal circumstances, reported one regimental commander, it was possible to advance from basic to First-Class pilot in as little as two years. See Colonel V. Kudryavtsev, "With Whom Are We to Fly into Combat?" *Krasnaia zvezda*, March 3, 1987.

when this problem first manifested itself even before the USSR's demise: "The prospects are unenviable—to go once again through all the wickets and retake examinations to qualify for ratings they previously held. This is simply an erosion of combat training."[53]

FLIGHT SIMULATOR UTILIZATION

Maintenance and planning days are also used to give pilots needed simulator practice if the regiment has a simulator that is up and functioning. Full mission simulators, if they exist at all, are rare in the VVS, and those equipped with an all-around visual capability and a complete spectrum of weapons employment options are most likely available only at weapons and tactics centers like Vladimirovka and Lipetsk. Most units are limited to rudimentary procedures trainers, much like the F-4 simulator which the USAF operated for years.

Simulator training is conducted haphazardly in the VVS and is underutilized to a point where pilots were said in at least one instance to have reacted with near-panic when informed by an inspection team that they would be tested for partial-panel instrument proficiency with the attitude director indicator (ADI) disabled. New pilots routinely make mistakes, often basic ones, in instrument flying technique. The VVS's monthly magazine, for example, reported the case of one upgrader who, given a partial-panel situation during a simulator check, rolled out onto final approach too far out and on the wrong heading.

In fairness, Russian pilots have voiced legitimate gripes about their simulators, the reliability and design of which are said to "leave a great deal to be desired."[54] One experienced pilot noted that the VVS's simulators were not only not helpful, but indeed imparted *negative* training because of their design imperfections. Another common complaint has focused on the frequent failure by regimental training supervisors to fully harmonize and integrate simulator training with the unit's flight training program.

---

[53]Lieutenant Colonel V. Usoltsev, "Victory in the Air and Problems on the Ground," *Krasnaia zvezda*, January 28, 1990.

[54]Colonel N. Gostev, "Perestroika in Training: What Is It to Be Like?" *Aviatsiia i kosmonavtika*, No. 1, January 1988, pp. 28–30.

A shortage of up-to-date simulators has long been a bane of VVS training, particularly for units converting to new equipment. Aircrews will frequently check out in a new aircraft type without the benefit of a simulator because of excessively slow construction of simulator buildings and the lengthy start-up times and delayed adjustment work caused by indifferent technical support from the manufacturer. Underutilization of simulators in line units (often as little as three hours a day) is not uncommon. Commanders often encourage inflated reporting of simulator use, logging as actual "flying time" the entire operating time of their simulators from morning start-up to evening shutdown.

WEAPONS TRAINING AND RANGE USE

The VVS operates scorable ranges for gunnery, rocketry, and practice bombing, as well as tactical ranges with more complex target arrays. According to Zuyev, these local *poligony*, as they are called in Russian, are situated close enough to most bases that aircrews can fly as many as three or four day range missions and two at night during mission employment training.[55]

These range facilities are not up to Western standards. Technical backwardness and a shortage of funds have required almost all range installations to be built locally by hand. In one reported case, a night bombing circle was illuminated by the placement of flat dishes of burning diesel fuel around the target, which required constant refilling and relighting. Aggressive scrounging of copper cable and the private purchase of light fixtures and bulbs enabled the installation of jury-rigged electrical lighting at one range.

Scoring systems are also primitive and typically involve a range officer calling out eyeball assessments. The fudging of scores is common, with figures padded more or less "depending on the rank of those who did the shooting." Critics of such practices have spoken out forcefully against "His Majesty the average rating, which ignores the obvious fact that the target is either destroyed or it isn't."[56] Such

---

[55]Zuyev, *Fulcrum*, p. 114.

[56]Lieutenant Colonel I. Kovalenko, "It Doesn't Seem Possible to Close the Ranges," *Aviatsiia i kosmonavtika*, No. 5, May 1991, pp. 28–30.

critics have noted that automated scoring systems could easily eliminate such abuses. However, the search for an enterprise in the Ministry of Defense Industry that might take on such work has failed for lack of funds.

Colonel General Shaposhnikov was a frequent critic of the VVS's failure to make the most of its of range facilities. "Year in and year out," he once observed, "pilots attack one and the same target, from sector to sector, on the same unchanging heading." Up to a point, he conceded, this was an unavoidable function of airspace restrictions or the geographic setting of the range. Yet more often than not, it mainly reflected "conservatism and a lack of desire among some commanders to think and to alter their routine."[57]

In addition to their day-to-day training at local ranges, VVS regiments during happier days periodically took part in surge exercises and weapons deployments. Fighter units flew one or two regiment-level tactical air exercises and five or six squadron-level exercises a year. During the early 1980s, the Soviet Air Force conducted periodic day and night airfield attack exercises involving squadron- and regiment-sized force packages operating out of forward deployment locations. The number of such exercises in 1983 reportedly exceeded the combined total for the preceding three years.

One 1983 exercise featured multievent air-to-air and ground attack operations, including fighter escort, low-altitude airfield attack, and independent search. Another featured operations against a high-value target, including extensive use of electronic countermeasures (ECM). By one informed account, it showed "a capability for coordinated attack not previously demonstrated by the Soviets."[58] Nevertheless, such mission employment scenarios remained highly canned and simplistic compared with what the USAF was doing in Red Flag and what the U.S. Navy was concurrently doing at NAS

---

[57]Shaposhnikov, "Devalued By Time," pp. 4–5.

[58]Captain Rana J. Pennington, USAF, "The Soviet Ability to Execute an Air Operation," unpublished presentation to a conference organized by the German Strategy Forum, Bonn, West Germany, 1984, p. 4. For a well-informed discussion of the concept of operations underlying such exercises, see Philip A. Petersen and Major John R. Clark, "Soviet Air and Antiair Operations," *Air University Review*, No. 3, March–April 1985, pp. 36–54.

Fallon. Dissimilar air combat training was typically limited to radar intercepts against nonmaneuvering targets.

In another exercise involving a Su-24 regiment tasked with simulating maritime attack out of a forward operating location, last-minute changes were introduced into the strike scenario, necessitating mission preparations under tight time constraints. A squadron commander authorized his aircrews to brief and fly a low-level attack profile, notwithstanding the fact that they "had not mastered this bombing technique before," because he had read about its successful application by Soviet airmen during World War II! The predictably poor outcome prompted a recommendation that each squadron maintain dedicated aircrews expressly trained for "out-of-the-ordinary" tasking. A similar recommendation for a building-block approach suggested that "after outstanding results are achieved in one mission category, effort should be shifted to assimilating the next tactical application." As matters stood, wrote the author, "the conveyor-belt method of training, in which aircrews go through the KBP with their afterburners on, so to speak, cannot . . . produce convincing evidence of improved proficiency."[59]

Before the disintegration of the USSR, the VVS's master operational air-to-air evaluation center was the Mary range complex adjacent to the Caspian Sea in the Kara Kum desert of Turkmenistan. Units would deploy there periodically for large-force employment operations and live weapons firing. The complex featured two airfields, one of which—Mary Two—was a full-time Frontal Aviation fighter base. Mary One was more built-up and diversified, with weapons and other depots to service a variety of combat aircraft types.

Unlike the USAF's practice at Red Flag, an entire regiment would deploy to Mary, from the commander through all aircrews and maintenance personnel to clerks, cooks, and drivers—and even the civilian waitresses in the officer's dining hall. The intent was to simulate a unit rotation to the Warsaw Pact forward area during a prelude to war. Deployments began with a formal in-brief of all regiment aircrews with the evaluation center staff. During the deployment, a unit could expect to receive a no-notice instruction to

---

[59]Colonel A. Tareyev, "An Operational Evaluation Over the Sea," *Aviatsiia i kosmonavtika*, No. 11, November 1987, pp. 10–11.

perform one or another of its most critical wartime tasks, much as in a USAF ORI. Acceptable scores for a regimental evaluation at Mary were 5/5 ("excellent") and 4/5 ("good").[60]

Toward the end of a two-week deployment, the unit would fly a regiment-sized operation against an equally large assemblage of aircraft simulating an enemy force package. There was no incorporation, however, of resultant operational "lessons learned" into the unit's subsequent training following its return home from Mary. There was also little or no tactical interaction or other cross-talk between the various units deployed to Mary for weapons-delivery evaluation.[61]

During his short tenure as commander in chief, Shaposhnikov indicated that because of budget cuts and Moscow's shift toward a defensive doctrine, the number of such annual exercises by the VVS was being reduced. Because of that, he said, a new role was being assigned to them as an "innovation laboratory."[62] Shaposhnikov explained that he wanted to separate these tactics development exercises from the ORI process and make them pure learning opportunities, seemingly on the order of Red Flag and comparable exercises conducted by the USAF.

A shortage of available weapons ranges had begun to develop even during the late Soviet period, when the number operated by the VVS (including those in Eastern Europe) dropped by a third following the withdrawal of the Western Group of Forces from the Warsaw Pact forward area. Further exacerbating the problem, the gradual liberalization of Soviet life under Gorbachev opened the gates for a profusion of citizen noise complaints and charges that the VVS's weapons ranges were public nuisances that ought to be shut down. The resulting "range starvation," in the expression of one lieutenant colonel, meant that opportunities to practice weapons delivery were becoming "more and more a rare holiday for pilots."[63]

---

[60]Zuyev, *Fulcrum*, pp. 190–203.

[61]Conversation with a former Soviet fighter pilot.

[62]Lieutenant General Ye. Shaposhnikov, "Toward a High Level of Combat Readiness—Through Concern for Others," *Aviatsiia i kosmonavtika*, No. 7, June 1989, pp. 1–3.

[63]Kovalenko, *Aviatsiia i kosmonavtika*, No. 5, May 1991, pp. 28–30. In contradiction to this, a knowledgeable Russian told me in 1993 that the loss of the Mary range complex

Today, weapons training in the VVS is almost wholly in remission because of the near-collapse of state funding to all the services for operations and support. According to the commander of Frontal Aviation, Colonel General Nikolai Antoshkin, the Soviet Air Force operated 80 weapons ranges, most of which were approved for live drops. The majority of these, including the Polesskii range in Belarus, the Mary complex in Turkmenistan, and a missile test range in Kazakhstan, were lost to the newly independent states when the USSR collapsed. The VVS now maintains only 36 ranges, 20 of which were set up solely to support rudimentary ground-attack training by the VVAULs.[64]

One measure of the VVS's determination to show that it remains operationally robust in the face of its continuing economic crisis, however, was the *Voskhod* '93 exercise conducted on May 18–19, 1993, under General Deinekin's personal supervision. That exercise was portrayed as an attempt to "simulate problems associated with the intratheater maneuvering of men, equipment, and weapons and with interaction between various air components during operations."[65] The stated intent was to fuse the actions of staffs, command posts, and airfields and to monitor activities spread over a great distance with no major failures.

The main deployment package consisted of six Tu-95 bombers, ten Su-24s, and four Su-27 escorts. These were supported by a dozen Il-78 tankers, one A-50 AWACS, and two airborne command posts for General Deinekin and his battle staff. The aircraft took off from three airfields in European Russia at 0100 local on May 18 and proceeded toward the Far East. The Su-24s conducted two in-flight refuelings

---

to Turkmenistan following the USSR's breakup would not significantly hurt KFA mission employment training, since there are enough range facilities in Russia to provide adequate weapons training opportunities for the considerably smaller Russian VVS.

[64]Interview with Colonel General N. Antoshkin, "Was There No Way the Missile Could Have Hit a Nuclear Power Station?" *Trud*, April 5, 1995.

[65]Report by Aleksandr Manushkin, "A Large-Scale Exercise Begins in the Russian Air Force," *Krasnaia zvezda*, May 18, 1993.

during a 12.5-hour flight, with the Su-27s landing and refueling at intermediate airfields.[66]

This deployment and the subsequent mission events performed at the Amur region test ranges were later described as something that "has no counterpart in the past decade" in Soviet or Russian practice in terms of complexity.[67] General Deinekin remained aboard an Il-62 airborne command and control center throughout the exercise to monitor its progress, with his headquarters staff chief, Colonel General Malyukov, handling ground coordination from the VVS's central command post in Moscow.

The deployment covered a distance of some 5000 miles and reportedly demonstrated good coordination, plus the ability of VVS aircrews to shift into a combat mode immediately after a long flight. The weather at the test ranges was bad enough to keep the Su-24s on the ground for a time. Their aircrews later performed low-level attacks against simulated targets. The two-day exercise ended with all aircraft recovering to their home bases at 1700 local on May 19.

Several explanations may account for this exercise. First, it showed a VVS commitment to make good on the emerging Russian military doctrine emphasizing rapid deployment of combat power to remote spots along Russia's periphery. Second, it may have sought to demonstrate, both to its own demoralized officers and to others, that the VVS had not lost its fighting edge despite its severe funding predicament. Finally, it may have been seen by the VVS as presenting a chance to exercise a new type of training so as to broaden its repertoire and identify hidden problems and weaknesses. What it lacked was an appreciation that the VVS's asset of greatest note for post-Soviet Russia's security needs was not high-performance strike

---

[66]Colonel Aleksandr Manushkin and Major Viktor Beltsov, "A Dash Across the Ural Mountains: The First Part of a Tactical Exercise Is Completed Successfully," *Krasnaia zvezda*, May 20, 1993.

[67]Colonel Aleksandr Novikov, senior navigator at the VVS Central Command Post, and Colonel Aleksandr Andryushkov, "*Voskhod* '93: The Command Post Is the Nerve Center of the Exercise," *Krasnaia zvezda*, May 21, 1993. During the early 1980s, the Soviet Air Force conducted squadron and regimental surges and demonstrated the ability of a fighter regiment to mount three separate regiment-sized attacks within a six-hour period. These activities did not, however, involve anything like the extended-range deployment featured in *Voskhod '93*. See Pennington, "The Soviet Ability to Execute an Air Operation."

aviation of the sort exercised in this deployment, but rather the less glamorous projection and sustainment of a ground presence through military airlift, as was painfully attested to by Moscow's involvement in Chechnya two years later.

TACTICS DEVELOPMENT

VVS fighter tactics, like Soviet force employment plans across the board, were considered state secrets throughout the long years of the cold war. Even today, they are not a topic of discussion in the more relaxed Russian military literature. Nevertheless, since the crumbling of many former security barriers that accompanied the disintegration of the Soviet state, some interesting insights have been provided by former Soviet pilots into the way in which the VVS pursued tactics development and application.

The home of VVS tactics development is the Lipetsk Center for Combat Readiness. There the KBP is written and periodically updated by a permanent cadre of instructors. Advanced tactics are also developed there for promulgation to line units.[68] Tactics documents developed at Lipetsk are endorsed by the Combat Training Directorate at VVS headquarters. The "new" tactics thus passed along, however, are often ignored down the line, with little change over time in the actual pattern of a unit's training cycle.

Air-to-air tactics are formalized in a VVS document called "Tactical Devices for Air Combat." Among other things, it describes named offensive and defensive maneuvers such as the "knot," "fork," and "mussel."[69] The tactics document also prescribes a comprehensive set of moves and countermoves for single aircraft and larger formations aimed at exploiting the full range of the weapon system's ca-

[68]Lipetsk accommodates the VVS's closest approximation to the USAF Weapons Center at Nellis AFB. It has long been a weapons training complex aimed at tactics development and validation for new equipment, syllabus development and manual writing, and upgrade training on new equipment for aircrews. For a snapshot overview of its mission and its current tribulations occasioned by the funding crisis, see the interview with its commander, Major General Nikolai Chaga, "On a Wing and a Promise," *Pravda*, January 25, 1995. See also Alexander Velovich, "Preparing for Combat," *Flight International*, February 14–18, 1995, pp. 26–27.

[69]Colonel Ye. Kulikovskii, "If Approached Creatively," *Aviatsiia i kosmonavtika*, No. 7, July 1992, pp. 4–5.

pability. This repertoire includes countering or negating anticipated enemy tactics through the use of maneuver, fire, jamming, and other measures.

Air-to-air tactics are divided into four categories:

- *Offensive*, for achieving a firing solution against an enemy from a tactically advantageous position
- *Defensive*, to negate an attack
- *Defensive-offensive* (or counteroffensive), to defeat an attack and regain the initiative
- *Neutral*, a set of positioning maneuvers intended to create a situation in which other friendly fighters can bring offensive tactics to bear.

VVS tactics development is a four-step process:

- First, operational threat assessment
- Second, diagramming an enemy's likely force employment, followed by a means of achieving surprise in breaking up his plans
- Third, scoping out the area in which initial moves will take place, with due allowances for one's own and the enemy's SAM dispositions and C3I nodes, as well as tactically significant weather, sun position, topographical features, and so on
- Fourth, deciding on appropriate formations and spreads, attack tactics, desired weapon configurations, use of electronic warfare support, and a disengagement plan that anticipates possible enemy counters.[70]

It was long suspected by Western analysts that there was a pronounced disconnect between what the Soviets preached in their tactics manuals and what they actually practiced in day-to-day training. That suspicion has since been borne out by subsequent revela-

---

[70]The article by Colonel Kulikovskii that provided this outline suggests that tactics should be gamed out on the ground before being employed in the air, using computer simulations and a full rehearsal of anticipated air combat. Unless this refers to new tactics development and validation, it indicates that even today, VVS tactical thinking and planning remain highly scripted in comparison with Western practice.

tions from Russian pilots and commanders. Lieutenant General Bobrovskii and Colonel Shubin, for example, noted that tactics development has "not undergone any significant change, even with the arrival of the latest aircraft into operational service." They also revealed that too many commanders concern themselves with tactical matters "only sporadically." Finally, they cited an absence of close and regular interaction between operators and the VVS's research and educational institutions—a deficiency that necessarily "limits the possibilities for integrating new tactical insights into combat training."[71]

Although each line unit has a tactics officer who attends a four-month qualifying course at Lipetsk, he gets little direct guidance from above, since of the many documents that pour into VVS units daily, "virtually none address the development of new tactical techniques." To fill the gap, said Bobrovskii and Shubin, squadron pilots develop their own tactical concepts, sometimes borrowing from squadrons with different aircraft types and amending them to reflect the capabilities and limitations of their own equipment. There was evidently no mechanism, however, for disseminating these "unofficial" tactics beyond the unit and integrating them into an overall VVS mission employment repertoire, since it was most often the case that "pilots in other regiments, unfortunately, do not know of them."

In a regiment that was among the first to convert to the MiG-29, tactics development was done in a "tactical theory class" in which pilots were given several situations and then directed to come up with the optimal solution. The most promising results were incorporated into the flight operations schedule and validated in the air. Then the most effective of these would be distilled and disseminated to all pilots.[72] Squadron and regimental instructors encouraged line pilots to decide for themselves what to do in "nonstandard situations."

Tactics developments within Frontal Aviation during the Soviet period were spread largely via word of mouth. Little information was

---

[71]Lieutenant General A. Bobrovskii and Colonel V. Shubin, "Tactics in Air Combat Training," *Aviatsiia i kosmonavtika*, No. 1, January 1990, pp. 1–3.

[72]Colonel I. Vokhubov, "A Direct Relationship," *Aviatsiia i kosmonavtika*, No. 10, October 1988, pp. 16–17.

made available to line aircrews on Western tactics. What was accessible was usually passed on to pilots during their conversion to new aircraft. There were no dedicated advanced training programs like the USAF Weapons School at Nellis AFB, the Navy Fighter Weapons School at NAS Miramar, or the Navy's Strike Leader Attack Training Syllabus (SLATS) at NAS Fallon. Tactics conferences for line aircrews were almost unheard of.

Bobrovskii and Shubin complained of a common situation in which mission plans look elaborate on paper, yet actual training sorties are flown "according to a time-worn scheme, without active opposition by an 'adversary.' And if there *is* opposition, the opposing force as a rule flies the same type of aircraft, and all moves are known beforehand right down to the slightest details." Zuyev later confirmed the absence of dissimilar air combat training as "one of the weakest points in Soviet training."[73]

The best VVS instructors recognize that effective fighter employment is not just a matter of good flying skills, but of clever tactics and anticipatory thinking aimed at defeating the enemy through guile. They also understand that tactics manuals can never provide "specific recommendations for every sudden twist of combat."[74] Even when breakthroughs were achieved at the margins in getting restrictions waived or regulations made less onerous, all too often there was no fuel or live ordnance available to permit trying ideas out. As a result, said VVS critics, much of the innovative thinking that leads to new concepts "remains on paper."[75]

The ten-year war experience in Afghanistan was a realism laboratory par excellence for Soviet fighter aviation. Inculcation of an operational mindset among VVS pilots began soon after a unit was informed that it would be deploying for combat. During initial workups of a squadron in Zuyev's MiG-23 regiment prior to deploying to Afghanistan, the aggressive commander declared frankly to his pilots that "safety will always get in the way of combat training.

---

[73]Zuyev, *Fulcrum*, p. 24.

[74]Lieutenant Colonel V. Drugovenko, "On Stratagem," *Aviatsiia i kosmonavtika*, No. 12, December 1987, pp. 12–13.

[75]Lieutenant Colonel V. Korotovskii and Captain S. Frolov, "A Hobbled Initiative," *Krasnaia zvezda*, March 20, 1988.

We're going to turn things around, and to hell with the safety inspector." The unit pressed ahead with its training as needed, often exceeding its weather minimums. But the commander reassured his pilots that there were no safety inspectors at the forward operating bases in Afghanistan.[76]

During their training workups prior to deploying, unit pilots flew multiple 20-minute range sorties every day. In Afghanistan, they averaged 300 combat missions during their year-long tour, often flying as many as four combat sorties in a single day.[77] Pilots on strip alert would often be assigned target coordinates and be airborne within ten minutes, routinely carrying out their mission despite the often tightly compressed time window within which to plan and prepare. Such success, said one returned pilot, was a natural outgrowth of the predominance of the mission over form and procedure.

Those who returned were different—seasoned by war, operationally mature, and no longer disposed to honor many of the canned procedures they had been brought up on in peacetime. An abiding attribute brought home by these combat-hardened pilots was a cultivated disdain for the bureaucratic routines that had previously dominated their professional lives. These pilots were beyond illusion, having experienced at first hand the virtues of flexibility in operations and tactics. They were indifferent to the old peacetime rules and harbored a case-hardened attitude toward their profession and its real-world imperatives.

One deputy squadron commander who flew 300 combat missions during his thirteen months in theater said that he had observed fewer mistakes, mishaps, and situations conducive to accidents while in Afghanistan than he could ever remember during peacetime. Why?

---

[76]Zuyev reported that at one point, the VVS sent groups of replacement pilots to Afghanistan only six months out of flight school, with barely more than a brand-new Third-Class rating. Said one unit commander: "Putting young boys like that into the cockpit of a Su-25 and sending them against Stingers is like sending sheep to the slaughterhouse." Zuyev, *Fulcrum*, p. 121, 172.

[77]Many returning pilots harbored stark memories of the Spartan living conditions at their bare bases. Asked of his most lasting recollection, one young captain said: "Plain *kasha* [buckwheat porridge]! I ate enough to last me the rest of my life! That's all we had for months." See Galina Marchenko, "Encounters at Airfields," *Aviatsiia i kosmonavtika*, No. 3, March 1990, pp. 28–30.

"I later understood that they trusted us, didn't harass us for trifles, and didn't foist office instructions upon us." By contrast, in the peacetime VVS, "they think more about honoring guidelines than about the job at hand. Here [back in the USSR], I'm completely at the mercy of restrictions." He proposed establishing an experimental regiment stripped of unnecessary rules and paperwork, just to prove what professionals could accomplish if left alone to focus on the mission: "It was tough in Afghanistan. But I felt myself a pilot there. There I was flying, whereas here I'm just meeting arbitrary requirements."

This battle-hardened cynicism and disdain for headquarters types (referred to dismissively by Zuyev as "staff rats") naturally unnerved many VVS leaders, who were concerned that the jaded outlook and disrespect for authority harbored by these seasoned war veterans might infect the remainder of the VVS.[78] Accordingly, rather than be used to pass along valuable combat experience to line squadrons, fighter units returning from the war were disbanded and their pilots spread randomly throughout the VVS to prevent the formation of clusters of rebelliousness.[79] For much the same reason, there was never any systematic effort by the VVS to identify and assimilate operational lessons acquired from the war experience.[80]

---

[78]Zuyev, *Fulcrum*, p. 120.

[79]Conversation with a former Soviet fighter pilot.

[80]For more on this, see Dennis Marshall-Hasdell, *Soviet Military Reform and the Afghan Experience: Military Lessons*, Conflict Studies Research Center, Royal Military Academy Sandhurst, Camberley, England, December 1993, pp. 37–40.

Chapter Six

THE OPERATING MILIEU

Until the last days of the Soviet Air Force, paperwork dominated daily life at the unit level. The regimental commander was constantly bombarded with documents containing instructions seemingly for every conceivable circumstance of life. The resultant profusion of red tape and the petty micromanagement that usually accompanied it (called *kantselyarshchina,* in colloquial Russian) reflected an innate distrust by VVS leaders in the professionalism and competence of their subordinates. The result was a stifling work environment fostered by the encrusted command-administrative system of governance which for years dominated Soviet military life across the board.

We know now from first-hand accounts by those Russian pilots bold enough to complain openly about it that the many rules and reporting requirements imposed on regimental commanders by higher headquarters were such that if a commander followed each restriction to the letter, his unit would be unable to generate a single sortie and remain legal. This forced regimental commanders to observe a double standard and, in effect, to live a lie on a daily basis. They would assiduously pay lip service to the rules in their reporting to higher headquarters, yet implement them only as they saw appropriate to support their training needs. The shameless cynicism and dishonesty of this practice reconfirmed the enduring relevance of the well-known adage by the 19th-century Russian satirist, Mikhail Soltykov-Shchedrin, that "the severity of Russian laws is softened by the nonnecessity of their fulfillment."

A VVS flying unit provided an unsheltering environment for officer career development. For those pilots who truly loved flying, a fighter squadron offered the best job in the country and the worst imaginable bureaucratic environment in which to perform it. Arbitrariness on the part of commanders was common, with little concern for people and their needs. A narrow focus on numbers and "looking good" forced constant compromises of integrity. Individual initiative was suppressed lest it result in a black mark on the unit. Cronyism was rampant, as was a widespread emphasis on the wrong priorities. Typically, mission readiness took a back seat to these considerations. With the lifting of many of the former Soviet restraints on free expression, thoughtful officers have owned up to these skewed values and acknowledged a serious need to fix the system. It is too soon to tell how successful they will be.

A TYRANNY OF BUREAUCRATISM

The Soviet pilot was always under a microscope, constantly taking written examinations, being subjected to monitoring on the ground and in the air, and having his mistakes broadcast openly by his squadron and regimental supervisors in a humiliating way for all to hear during group *samokritika* (self-criticism) sessions.[1] He was incessantly beseeched by multiple bosses to "show initiative." Yet he knew that beneath it all was the cardinal unspoken rule: "Don't screw up!" Particularly when it came to losing or damaging airplanes, the VVS was a zero-mistake air force.

One instructor bemoaned the "patent absurdity" of the way in which continuation training was organized in line units. The KBP, which he called the "fundamental document" for training, was smothered by a bureaucratic approach that fostered "an absence of normal thinking and initiative" and "a system that sees to the welfare of careerists, the ungifted, and the play-it-safe crowd," who together obstruct the realization of the performance goals laid out in the KBP. Constantly hovering over the squadron commander was such a "knot of directives" levying upon him so many requirements from higher headquarters that he ended up being "tied hand and foot just reading

---

[1]Captain Yu. Zhukovskii, "The Price of Carelessness," *Aviatsiia i kosmonavtika*, No. 11, November 1987, pp. 28–29.

them." On top of that, a squadron commander would frequently work late into the night with his deputies building a sensible flying schedule for the next day, only to have the phone ring from higher headquarters with some supervisor announcing: "I'll be flying with you tomorrow, so plan this many flights." The predictable result was to send everyone back to square one "with cursing and nervous strain."

Military Transport Aviation (VTA) was no less afflicted by bureaucratism than the fighter and bomber communities. A 30-year VTA veteran railed against the proliferation of documents and forms which, he said, "swamped everyone from pilots to maintenance technicians, and especially commanders. Ask any of them what hamstrings them the most. Without hesitation, they will single out the paperwork. It corrupts the operational sense of pilots, dulls their memory, and breeds lethargy. Wags even joke about it—'the scratching of pens drowns out the whine of turbines.'"[2]

This critic, a retired lieutenant colonel, went on to complain that "all preparation for flying . . . is pervaded with formalism and is reduced essentially to filling out forms in the interest of management oversight that, at best, has little bearing on the performance of the mission." Such directives, he grumbled, "are more suited to a robot than an intelligent human." He further faulted the prevalence of cronyism in VTA, which meant that "people whose qualifications lie well below those of their subordinates are nevertheless sometimes picked to fill supervisory positions. They are not selected on their merits, but rather as a result of patronage from above."[3]

Career prospects were not routinely bright for senior captains and majors hoping to gain positions of command and higher leadership.

---

[2]Lieutenant Colonel (Res.) V. Shishkin, "A Hope Undimmed By Years," *Aviatsiia i kosmonavtika*, No. 12, December 1991, pp. 6–7.

[3]In an interesting side comment, he further noted that even with digital computers now installed in most current-generation aircraft, aircrews remain forced to do their mission planning using the venerable NL-10M wooden plotter. He granted that headquarters recognized the problem but typically countered with the lame excuse that "there is no funding for the *Luch*-84 computer." To this, the colonel wryly commented: "One can, of course, also drive nails with an electric iron and crack nuts with a crystal vase. But wouldn't it make more sense to use these items, and the system, for their intended purposes?"

For one thing, advancement in rank, even for lieutenants, required an available billet appropriate to the next level, which explains why there were so many 40-year old captains and majors in line regiments. Promotion required not just satisfactory performance of one's duties but patronage.

Also, unlike USAF pilots, who typically rotate from one assignment to the next every three years, Russuan pilots might remain with the same unit, and often at the same base, for most of their careers unless they expressly request a transfer. This has made the aircrew composition of a VVS regiment more like that of an American Air National Guard or Air Force Reserve unit in terms of manning stability over time. The positive side is that this has allowed, at least in principle, for the development of a closely knit unit. The negative side is that because the influx of new blood was typically limited to new RTU graduates coming in to replace older pilots retiring from flight status, units easily tend toward complacency and stagnation.

During the 1980s, it was not uncommon for VVS headquarters to force experienced pilots into early retirement to make room for new flight school graduates. One officer who flew in Afghanistan and later retired as a major general sharply criticized this practice of summarily declaring pilots with eight to ten years of cockpit time "old" and lacking a future when, in fact, they represented the core of the VVS's talent pool. He charged that it was senseless to invest a million rubles toward producing an experienced military pilot, only to turn him out at the prime of his proficiency so that the same amount might be reinvested in generating new pilots.[4]

Cronyism was a major factor in selection for command. It was widely known at the VVS working level, complained one officer, that "the fates of all commanders are decided in the nether reaches of the personnel world. And it is no secret that family ties or the backing of influential people, sometimes with no connection whatever to aviation, can have a determining significance. The secrecy associated with the selection of regiment commanders is one of the main reasons for the encroachment of mediocrity up the service hierarchy."

---

[4]Colonel Aleksandr Tsalko, "We Do Not Value Thrift," *Krasnaia zvezda*, January 24, 1989.

By almost unanimous testimony of both defectors and those who stayed to criticize the system from within, arbitrariness was the rule governing peacetime VVS training. Typically it took years for a pilot to attain First-Class status, often due to nothing more than the needlessly lethargic pace of a commander's training program. Alexander Zuyev pointed out how a VVS ruling that all new pilots would henceforth be expected to achieve Third-Class status within a year forced his regimental commander to work harder than he would have otherwise. This underscored the near-total absence of homogeneity in the way VVS fighter units were run.[5] Zuyev further cited the case of a regimental commander who would routinely cancel flying during "dangerous" weather, leaving it to himself to determine arbitrarily what constituted "dangerous." This commander was unabashed in acknowledging that his only interest was avoiding the loss of any aircraft, even if that meant ignoring the needs of readiness altogether. Indeed, he went so far as to proclaim that "the less you fly, the longer you keep on flying."[6]

Safety almost invariably took precedence over realism in mission employment training. Zuyev told a story about his RTU commander who tried to set an example during workups to a division weapons meet by leading his pilots in trying out a new maneuver to defeat the American-made Stinger shoulder-fired infrared missile at the weapons range, when the safety supervisor came up on the radio, shouting: "Stop this gross violation! Stop this hooliganism immediately!" The squadron commander got chewed out afterward but was not seriously reprimanded. Zuyev remarked that his innovative maneuver might have saved lives in Afghanistan, but that it was never added to the combat training syllabus.[7]

---

[5]Alexander Zuyev, with Malcolm McConnell, *Fulcrum*, New York, Warner Books, Inc., 1992, p. 114.

[6]Ibid., p. 111.

[7]Ibid., p. 129. The introduction of the Stinger in Afghanistan in October 1986 had such an impact on the VVS's loss rate that Soviet strike aircraft pilots were forced to deliver their ordnance from high altitudes, seriously eroding their bombing accuracy. For a time, Mujaheddin rebels were claiming at least one Soviet combat aircraft downed each day, with a reported 65–70 percent kill rate per missile launch. See John Gunston, "Stingers Used by Afghan Rebels Stymie Soviet Air Force Tactics," *Aviation Week and Space Technology*, April 4, 1988, pp. 46–48.

There was likewise a pervasive emphasis on "looking good," even if that, too, impeded readiness. Pilots were repeatedly forced to perform so-called "vitally important housekeeping duties" at the expense of their primary responsibility for staying mission ready. No wonder, complained one, that aircrews were fed up with being upbraided more because some fence around the regiment's headquarters needed fixing than for a near–midair collision in air combat training. Such distortions of legitimate priorities were a major reason why so many pilots were electing to resign their commissions and leave the service for civilian life. Commanders at all levels, he said caustically, are paid to see to their unit's readiness, not to paint fences. Yet all too often, they default on this responsibility by "cooking the books" to look good, while squandering their remaining time on marginalia. "Reports above all!" is the main measure of a commander's merit, he charged, leading to the dominance of "formalism and efforts by commanders to meet their assigned goals at any cost, whatever the consequences."[8]

SELF-DECEPTION AS A WAY OF LIFE

A triumph of form over substance characterized VVS self-assessment throughout the many years of Soviet rule. In a case in point, one general admitted that the flying establishment had come to judge a pilot's skills in the cockpit "mainly by the number of badges on his chest—namely, on the basis of his class rating," when what truly mattered was his ability to hit ground and airborne targets accurately in various weather and tactical conditions.[9] He added that under the existing rating system, "everything is turned upside down. We make weather conditions the dominant criteria, while the ability to employ one's weapons, which reflects the true level of proficiency, is buried away in obscure columns on a mission evaluation form."

There were parts of the USSR, notably in Central Asia, where a pilot could *never* earn the top aeronautical rating, since the adverse weather conditions required for such a rating never occurred there.

---

[8]Captain S. Prokopenko, "Both Pilot and . . . Trackman," *Krasnaia zvezda*, April 8, 1989.

[9]Major General N. Posrednikov, "The Class and Proficiency Rating," *Krasnaia zvezda*, January 11, 1990.

This led to a serious lapse of integrity in which commanders would simply lie about the weather on the certification forms in order to allow a pilot to attain a First-Class rating, even though the required weather conditions were never met. In lieu of this misplaced focus on weather minimums, complained the VVS general, "the key measure of a military pilot's skill should be his ability to fly a combat mission. Airfield weather conditions . . . should be of only secondary importance."[10]

Worse yet, he added, the recurrent practice of dishonest reporting made for a guaranteed safety hazard, in that pilots could be cleared for flight in IFR situations for which they had never actually qualified. At the root of this problem was a rampant "percentomania" which impelled regimental commanders to accelerate the advancement of their pilots to First-Class status so their units might look good on the books, regardless of whether the pilots in question had actually accomplished the events and undergone the proficiency checks required for that aeronautical rating.[11]

Another such example of dishonesty cited during the early revelations of glasnost was a willful falsification of a mission write-up in the wake of an over-*g* incident during a recurrency flight in a dual-control fighter. The requalifying pilot was noncurrent to the point of not having performed advanced handling maneuvers for over two years. The scheduler knew this, yet raised no objections when the requalifying officer was approved to fly a sortie and perform events for which even the accompanying instructor was not qualified.[12]

After landing, the instructor failed to report the over-*g* incident. The whistle was nevertheless blown by the chief of the squadron's flight recorder section. The issue eventually percolated up to higher head-

---

[10]General Posrednikov argued against eliminating proficiency rating pay, which served as a stimulus for improving one's proficiency. He maintained, however, that the effectiveness of that stimulus could be enhanced "by rejecting leveling practices. The amount of the reward should depend directly on the number of hits and misses on range missions and on the complexity of training sorties and the skill with which they are carried out. Then no pilot will take off on a range mission with a defective sight."

[11]Colonel T. Sheshenya, "The Combat Nucleus of a Regiment," *Krasnaia zvezda*, June 15, 1988.

[12]Colonel V. Yudin, "Deformation," *Krasnaia zvezda*, May 5, 1988.

quarters, prompting a ruling that faulted the *requalifying pilot* for having introduced a control input that the instructor had been unable to countermand in time to prevent the over-g, rather than properly reprimanding the instructor instead for having allowed an unqualified pilot to exceed his limitations and then failing to report the event.

Lieutenant Viktor Belenko, the VPVO pilot who defected with his MiG-25 to Japan in 1976, described from personal experience a related instance of "cooking the books" years earlier, indicating that such self-deception had a long tradition in Soviet military aviation. During unit workups before a scheduled headquarters inspection, he was slated to fly a one-hour 2 v 1 intercept mission against an upgrading student in a MiG-17. On the morning of the scheduled event, local area thunderstorms threatened to cancel the mission. Belenko was nevertheless ordered by his deputy regimental commander to fly the mission, with the following assurance: "Listen to me. Just tell your student to climb up to 500 meters. You make a quick intercept, and both of you come right back down. It won't take five minutes. I'll show you how to fix it when you get back." For the next three days, Belenko and his deputy commander juggled gun camera film and flight recorder tapes to concoct a record of an elaborate and successful mission. To mask the discrepancy between the sixty minutes of reported flying time and six minutes of actual time aloft, the unconsumed fuel for all three aircraft—thousands of pounds—was dumped onto the ground.[13]

Even before his appointment as commander in chief, General Shaposhnikov freely acknowledged the overbureaucratization of the VVS that occasioned such unconscionable lapses in integrity. At one point, he remarked how "our entire service life is thoroughly regimented and regulated by the requirements of multiple orders, directives, and regulations."[14] One of his favorite targets was rampant micromanagement and the arrogation of all wisdom and decision-making power to the highest levels of command. Representative of

---

[13]John Barron, *MiG Pilot: The Final Escape of Lieutenant Belenko*, New York, McGraw-Hill Book Company, 1980, pp. 82–83.

[14]Lieutenant General Ye. Shaposhnikov, "Toward a High Level of Combat Readiness—Through Concern for Others," *Aviatsiia i kosmonavtika*, No. 7, July 1989, pp. 1–3.

this attitude was a remark he made while still air commander in the GSFG that "excessive supervision brings nothing but harm to training."[15]

Similar concerns were later expressed by VPVO's commander of fighter aviation, Colonel General Vladimir Andreyev. With remarkable understatement, he described how he had been fired by the VPVO commander in chief, General Tretyak, during the final year of Soviet rule because he had become "inconvenient" in drawing attention to rampant corruption and abuses. In a seminal comment on the importance of integrity, he stressed that "if you begin practicing deceit in aviation, the consequences will be tragic."[16] General Andreyev said that it had been his practice to highlight problems for the VPVO commander in chief as he saw them, drawing freely on the related experience of the VVS, naval aviation, and other air forces around the world. Yet despite this, he confessed, the views of the professionals in his fighter directorate were given no respect, as "General of the Army" Tretyak routinely issued contrary decisions: "He would hand down orders that caused line pilots to pull their hair out." Repeatedly, Andreyev was forced to protect his people from unjustified attacks by their out-of-touch commander in chief. Looking ahead, he concluded: "I don't want to discuss him any more. What matters is that there will be fewer such military leaders in the future."

THE BEGINNINGS OF SOVIET AWAKENING

The absence of realism in fighter training, long a repressed concern throughout the flying community, came under increasing attack from within the VVS's ranks once glasnost lifted the lid on open criticism of established practice. Although the Soviet fighter pilot was expected to be fully ready in case of war, "the sad thing," complained two officers in 1988, "is that we talk a lot but are doing far from every-

---

[15]Lieutenant General Ye. Shaposhnikov, "The Degree of Readiness," *Krasnaia zvezda*, June 8, 1988.

[16]Interview with then–Lieutenant General Vladimir I. Andreyev, "We Need to Know the Threat by Sight," *Krasnaia zvezda*, November 22, 1992. Andreyev was quickly reinstated following the abortive 1991 coup after General Tretyak was fired for backing the putschists.

thing possible to attend to this."[17] These writers went on to say that even today, "you would think we were doing anything but sitting on our hands, since the airfields know no silence day or night. Yet take a closer look. Little gain is being achieved. You will notice that here they played it safe, there they made things easy, and over there they overlooked things. As a result, we aren't getting the payoff that we should."[18]

These critics stopped short of a blanket indictment of VVS practice. Yet they were frank to insist that "some of the provisions [of the rules] are out of date, are geared toward training pilots under simplistic conditions, and fail to reflect current needs." They complained that "flights are standardized by official documents literally from takeoff to landing." As a result, they said, "the slightest deviation or initiative by a pilot is typically penalized as aerial hooliganism [the Soviet term for wanton disregard for air discipline]," even though it is common knowledge among pilots that "one can't do the job properly in combat training without intelligent and measured risk."[19]

Toward the end of the Soviet era, VVS pilots were openly demanding that demonstrated competence in advanced handling should be a prerequisite for more complex mission employment training. As one remarked, "the minimum [requirement] stipulated by the training plan must not be confused with adequacy. The minimum is *only* a minimum, and meeting it even 100 percent hardly constitutes grounds for rejoicing." Citing the well-known fact that an aviator skilled at air combat maneuvering should also be adept at less demanding mission events, he noted that he had "never encountered an outstanding aerobatic pilot who had difficulty with other forms of

---

[17]Lieutenant Colonel V. Korotovskii and Captain S. Frolov, "A Hobbled Initiative," *Krasnaia zvezda*, March 20, 1988.

[18]In 1989, an informed civilian defense specialist, Vitaly Shlykov, castigated the "Stalinist art of winning by numbers rather than by skill" that had been practiced by the Soviet armed forces in World War II. Shlykov contrasted poorly trained Soviet pilots (including, he said, leading aces Pokryshkin and Kozhedub) with the far superior aces of the Luftwaffe, who were "given proper training and used sparingly, being valued for their skill in aerial combat." Interview in *International Affairs*, No. 5, May 1989, p. 23.

[19]Korotovskii and Frolov, 1988.

flying," whereas "situations involving the reverse are not uncommon."

Similar complaints were voiced by VVS aircrews over the culture of dishonesty that senior echelons had allowed to predominate over VVS practice for years. They were well reflected in a February 1988 article lambasting the absence of integrity within units and the pervasive resort by commanders to the "whitewashing of shortcomings" and even outright cheating to score well in operational readiness inspections.[20]

The ultimate indictment of the VVS's shortcomings in training was the acid reply of a Soviet pilot who was asked whether he felt that the 1991 Persian Gulf War would have ended differently had Iraq's aircraft been flown by Soviet pilots. This officer replied: "Hardly, because Iraq's pilots were trained by our pilots." When pressed to affirm that "surely [Soviet] pilots are not that bad," the officer countered that "any thinking [Soviet] pilot today knows that in case of war, he is assigned the role of cannon fodder. He also knows that this bothers very few people at the top."[21]

In a headquarters response to these outcries of discontent, the VVS's deputy commander for combat training, Colonel General Anatoly Borsuk, kicked the ball back to subordinate commanders, stating that in so complaining, they were failing to exercise their vested command prerogatives. Rejecting the charge that all restrictions "flow from the central VVS staff and nowhere else," Borsuk maintained that although "some time ago such prohibitions did exist," under perestroika the responsibility for the most important prerogatives connected with aircrew training had "been delegated directly to regiments and squadrons." Suggesting that unit commanders should look to themselves rather than point the finger of blame at higher echelons, he insisted: "Today, all the guidance documents are oriented toward giving commanders the greatest possible indepen-

---

[20]Lieutenant Colonel G. Belostotskii, "Wait . . . A Test Target," *Krasnaia zvezda*, February 5, 1988.

[21]Interview with Lieutenant Colonel Vasily Vysotskii, "We Are Rarely in the Air . . .," *Komsomolskaia pravda*, August 7, 1991. A senior captain stated that the Soviet Air Force "could not have operated as crisply in the Persian Gulf as did the Americans. The overabundance of instructions does not yet allow us to reach their level."

dence."[22] Borsuk acknowledged the errors of the past, "when initiative could only flow from the top down." Yet he stressed that "under [current VVS] guidance, regimental commanders have been given every opportunity to decide independently how to conduct the training of their units' pilots."

With the gauntlet thus thrown down for operators to show their ability to exercise the new latitude that had been granted them from above, it was only a matter of time before bolder voices from the flying community would sound off in reply. One of the first came not from the VVS, but from the respected chief test pilot of the Mikoyan Design Bureau, Valery Menitskii. In a hard-hitting critique of the stolid conservatism that continued to afflict VVS training and tactics application, Menitskii cited from his own experience cases of pilots whose air combat prowess had been all but nonexistent from their first day of graduation from flight school, because of the "ban that existed at that time on advanced aerobatics." The root cause, he added, was not the pilots or their commanders but "the system itself. . . . The problem was that none of the top leaders were willing to face up to this during those 'stagnant' times."[23] Menitskii noted with dismay how seasoned test pilots on his staff would routinely visit line regiments to see how the users were doing with their new equipment, only to discover all too often "units where they downplay the role of flying skills and approach tactics in a cookbook manner."[24]

Menitskii chided the VVS for its reluctance to run legitimate risks in the interest of increased air combat proficiency. "True enough," he said, "safety can be achieved through bans. But such safety is illusory, producing unhappy results in the end. The bitter experience of

---

[22]Interview by Colonel V. Seledkin with Colonel General A. Borsuk, "To Assume Responsibility More Boldly," *Krasnaia zvezda*, July 12, 1988.

[23]Valery E. Menitskii, Honored Test Pilot of the USSR, "Flying Skills: The Key to Victory," *Krasnaia zvezda*, May 11, 1989.

[24]As a case in point, the chief test pilot of the Yakovlev Design Bureau remarked how some Mikoyan pilots had visited one MiG-29 regiment whose pilots were restricted to 6 *g* in the aircraft (it is designed for normal operations at up to 9.5 *g*): "And it is like that for many flight parameters. What kind of combat readiness and combat capability can you talk about when a pilot doesn't have any idea of his aircraft's potential?" Interview with Andrei Sinitsyn by Major S. Prokopenko, "The Yak-141: The Aircraft, People, and Problems," *Krasnaia zvezda*, August 17, 1991.

bans has more than once put military pilots in a bad situation." He added: "Almost all pilots understand this, but unfortunately, not all commanders share their view. They command regiments and squadrons as prisoners of old concepts. Such commanders oversimplify tasks in all sorts of ways, instruct subordinates to fly over the same routes and execute attacks using simple types of maneuvers, thereby seeking to reduce the number of accident-prone situations in the air. In so doing, they train pilots not to be creative."

Menitskii's complaint was echoed by a retired colonel with thirty years of fighter experience, who revealingly noted that once attention had become focused on air-to-air missile employment, "the training of pilots in close maneuvering air combat was halted. Such exercises were removed from the KBP. What did this lopsided arrangement lead to? The well-known Israeli-Egyptian military conflict revealed the unsoundness of the idea of arming the VVS only with long-range and short-range missiles." He added that this was not merely a point of quaint historical interest: "*At the present time, supporters of the elimination of close maneuver air combat from the arsenal of tactical methods of pilots are appearing once again* [emphasis added]. There is a need for serious study of tactics today."[25]

A related comment confirmed that although much had improved since the earlier days of stereotyped conduct, "the combat training system . . . remains unfortunately to this day grounded on the old foundation" and "its essentials have remained unchanged."[26] The co-authors of this complaint charged that existing VVS approaches to tactical training continued to "lag far behind the sophistication of contemporary aviation equipment," in considerable part because of the "voluminous instructions that fall plentifully on decrepit regiments 'from above.'" They warned that it would be no mean feat to change the existing state of affairs for the better, since the long-established VVS approach to aircrew training had occasioned "a dearth of initiative, creativity, and capacity to assume responsibility." They added that the prospects for any real revolution in combat training

---

[25]Colonel (Ret.) V. Uryuzhnikov, "Don't Make the Mistakes of the Past," *Kommunist vooruzhenykh sil*, No. 16, August 1989, pp. 13–15.

[26]Lieutenant Colonel of Medical Service V. Koslov and Lieutenant Colonel A. Zhilin, "The Pilot in the Combat Training System," *Aviatsiia i kosmonavtika*, No. 8, August 1990, pp. 10–12.

would "depend on how soon we begin ridding ourselves of our serf mentality."

In an indication of how far the VVS had yet to go in converting promise to reality, one squadron commander, whose pilots were actually trying to apply the new license granted by higher headquarters to experiment with exercising greater initiative, concluded that things were not working out. For years, he said, "we were forced into a situation in which combat training . . . amounted to a one-way road to mediocrity, which is, judging from all the evidence, where we've ended up."[27] Following this confirmation of a truth about Soviet training that had long been widely suspected in the West, the squadron commander added: "We learned to suppress any initiative during our many years of living under the thumb of the command-administrative system. We still are fearful of legalizing it altogether—what if everything that goes around today comes around again tomorrow?"

This officer recalled in frustration how he and his deputies had struggled unsuccessfully to escape the old mold by examining individual records and seeking to determine the actual proficiency of each pilot all over again: "Most notable about this was how it showed that we'd succumbed to the illusion that we knew each other. Fat chance! We didn't even know ourselves well, since that had not been particularly important before. The abstract image of some statistically average First-, Second-, and Third-Class pilot had been formulated automatically. The entire process of combat training had been geared toward statistically average mediocrity." This conscientious leader spoke enviously of having recently read an interview with a counterpart RAF fighter squadron commander, which had revealed a standard of excellence to which Soviet professionalism could still only aspire.

TOWARD NEW HORIZONS

During his brief incumbency as commander in chief, General Shaposhnikov laid the groundwork for a top-to-bottom revamping of

---

[27]Lieutenant Colonel A. Tokarenko, "Testing By Initiative," *Aviatsiia i kosmonavtika*, No. 12, December 1990, pp. 4–5.

VVS approaches to fighter operations and training. He clearly recognized the VVS's long-unrequited need for greater training realism, and he advocated innovations, alluded to in the preceding chapter, that sounded more than passingly similar to accepted Western practices. Examples include his suggestion that exercises should be geared more toward training and tactics development than to readiness certification, his insistence on strict standards for allocating sortie events among aircrews in squadron training, and his call for a yardstick by which Soviet pilot proficiency might be measured using an approach that sounded, in concept if not in form, roughly analogous to the USAF's Graduated Combat Capability (GCC) system.[28]

Finally, and most important, Shaposhnikov recognized the weakness emanating from the VVS's suppression of pilot initiative, as well as its overreliance on scripted sortie profiles under GCI close control. Soviet pilots had routinely grumbled about that self-inflicted limitation for years.[29] This was the first time, however, that a VVS commander had openly agreed and announced a determination to bring about needed changes. In earlier times of East-West tension, the rise of such an innovator to be commander in chief of the VVS would have been grounds for legitimate concern among NATO planners. Yet with the USSR well on its way toward imploding in December 1991, the most pressing question was whether Shaposhnikov would be able to overcome the bureaucratic drag, long-standing habit patterns, and deep political uncertainty that, together, threatened to obstruct the realization of his sought-after reforms.

Today, with the Soviet system repudiated and a new horizon looming ahead, the VVS under General Deinekin stands on the threshold of *potentially* the most radical departure from its familiar ways since the earliest days of the Soviet state. As the chief of VVS education and training commented before the USSR's final collapse, "the

---

[28]This is not to suggest that Shaposhnikov had in mind emulating USAF practices per se. It does, however, mean that he seemed interested in bringing about changes in VVS tactical training motivated by much the same sort of operationally oriented thinking that has characterized USAF training since the birth of Red Flag and DACT in the mid-1970s.

[29]In a typical venting of such frustration, one VVS captain wrote as early as 1975 that "it is all very well that GCI operators should assist us fighter pilots, but one should not rely on their support for everything." Captain A. Potemkin, "Respond to the Situation," *Aviatsiia i kosmonavtika*, No. 12, December 1975, p. 15.

events of August 1991 have accelerated the process of radical change in the country's armed forces. It is gratifying to note that common sense is returning to us, albeit slowly."[30]

This reflection underscores the important point that throughout the VVS's history, the main problem was the Soviet system, not the individual pilot or his equipment. The Soviet pilot was selected by exacting criteria, and he represented the best talent for his calling that Soviet society had to offer. Soviet aircraft and air-to-air missiles have always been respectable from a technical standpoint. Properly employed, the fourth-generation MiG-29 and Su-27, with their AA-10 Alamo and AA-11 Archer missiles, are a match for any comparable systems the West currently operates. In some respects, they command a decided performance edge. The improved MiG-29M, the Su-35, and a new Russian AMRAAM-equivalent missile now in advanced development promise further advantages yet. The reason the VVS has long had such trouble getting the most out of these assets is that the Soviet pilot was inevitably a product of his training environment. Naturally, his techniques and skills were heavily conditioned—and circumscribed—by the inhibiting influence of a uniquely "Soviet" operational culture.

Considering the many restrictions on pilot initiative that hampered the VVS's operational adaptability throughout the long years of the cold war, it was all but inevitable that the Soviet MiG-21 pilots who were lured into battle by the Israeli Air Force over Suez in July 1970 would have been so completely outmatched. According to first-hand accounts by the Israeli pilots who engaged them, the Soviets were aggressive and flew textbook formations going into the fight. Once the engagement was joined, however, their mutual support quickly broke down and they began making elementary mistakes, including indiscriminately firing their early-generation Atoll infrared missiles not only outside of effective parameters, but seemingly from panic and to no apparent tactical purpose. By the end of the five-minute melee, five MiG-21s were downed for no Israeli losses.

The difference was not in the quality of the individuals who were pitted against one another in that engagement, but rather in their dia-

---

[30]Major General Ya. Yanakov, "From the Retrograde to the Modern," *Aviatsiia i kosmonavtika*, No. 1, January 1992, pp. 2–3.

metrically opposed approaches to training and force employment. With their heavy dependence on GCI and their unfamiliarity with anything beyond the broadest essentials of free air-combat maneuvering, the Soviets lacked the situation awareness and implicit knowledge of appropriate moves and countermoves that are crucial for surviving and winning in a dynamic, multiparticipant air battle. That said, it is a safe bet that a typical Russian fighter pilot today could be picked virtually at random from squadron service, detrained of his most counterproductive habits acquired through exposure to Soviet influence, enrolled in a USAF F-15 RTU upgrade course or its equivalent, and emerge with creditable air-to-air skills by any standard.

Since the USSR's collapse, the VVS has been freed of the organizational chokehold that limited its capacity to innovate under Soviet rule. In principle, it is now at liberty to cast aside its old ways and develop a new operational repertoire aimed at extracting the fullest leverage from its highly capable equipment. Yet with a shoestring operations and support budget that forces commanders to bend every effort simply to maintain their pilots' basic aircraft handling proficiency and instrument and landing currency, it is hard to imagine how they might conduct the sort of training (like that set forth in the USAF's Multi-Command Manual 11 series) that would be required, at a minimum, to bring Russia's pilots up to accepted Western standards.

In this connection, Colonel General Borsuk's successor as deputy commander in chief for combat training, Colonel General Yevgeny Zarudnev, conceded in late 1992 that the state of training at the regiment, division, and air army level had become "catastrophic."[31] In contrast to the former Soviet ideal norm of 140–160 flying hours a year for fighter pilots, he reported that the VVS was registering only a quarter to a fifth of that because of inadequate funding. Zarudnev added that maintenance manning was down to 50–70 percent of normal levels in most units. He confirmed that VVS leadership had given "little thought" to the requirements for realistic training throughout the Soviet period and that it now faced the hard choice of

---

[31]Interview with Colonel General Ye. Zarudnev, "We Must Not Lose Control of the Situation," *Aviatsiia i kosmonavtika*, No. 12, December 1992, pp. 2–3.

whether, in the face of its pilot surplus, to let its youngest pilots go and thus sacrifice the next generation or to retire its more experienced pilots early, thus leaving the VVS with no significant combat capability. The extent of the VVS's troubles in this respect was graphically shown two years later by the repeatedly deficient performance of its ground attack pilots in the war in Chechnya as a result of chronically underfunded training (see Chapter Eight).

To recall a point noted at the beginning of this study, some of the afflictions highlighted by Russian airmen since the onset of glasnost are endemic to most air forces around the world, the USAF included. The one tendency that remains unique to the VVS, however, is the top-down rigidity in both operations and thought that the communist system, for years, imposed on pilots and commanders who knew better but were obliged to pretend otherwise. Because old habits die hard, elements of it persist to this day, even though flight activity has been reduced to a near-halt because of the budget crisis. It is this legacy of the now-discredited Soviet approach to operations and training that Russian airmen will have to work the hardest to overcome.

Chapter Seven

HOW MIGHT THEY HAVE DONE IF . . .?

With the cold war now over and the Soviet Union history, it is instructive to revisit the debate about Soviet air combat prowess that raged among Western analysts during the 1970s and 1980s, in search of better insights into the once-deadlocked question of how the VVS might *really* have acquitted itself in an aerial showdown over Central Europe. The point of such an exercise is not that we now have a chance to take a better look at the VVS's strengths and weaknesses against the possibility that a failure of reform in Russia will again pit us against our former rival. Despite Russia's current troubles and the likelihood that its future evolution will include reverses that will make for strained dealings between Washington and Moscow, it remains U.S. policy to bend every effort to engage Russia in a cooperative security relationship for the long haul.

Beyond that, the USAF over the past five years has conducted a mutually rewarding exchange relationship with its Russian sister service aimed at building bridges and slowly institutionalizing professional ties at all levels so that any reversion to confrontation will be that much more difficult. It would be counterproductive to send messages to the VVS leadership—and to the Russian defense establishment more generally—implying an unseemly interest in their training practices that might undermine this important but still-fragile relationship. Rather, the point of asking this question is to take advantage of fresh evidence like that brought together in the preceding chapters, to inform a retrospective look at where we were right, where we were wrong, and what methodological conclusions we can draw from such insights regarding mistakes to avoid in *future* efforts to come to grips with force employment practices that diverge from our own.

Early in the nuclear age, when the USSR remained all but opaque to outside scrutiny, Raymond Garthoff wrote insightfully that "in order to understand an alien military culture, it is first necessary to escape the confines of one's own implicit and unconscious strategic concept. The ideas of others, when these are interpreted in terms of the military or political analyst's *own* strategic preconceptions, will appear distorted or, often, obsolete. And the comfort derived from a superficial assessment of differing views, in such a manner that these views seem to represent a simple 'cultural lag' on the part of our opponent, may obscure the recognition of these views as manifestations of a different underlying doctrine and strategic concept."[1] At the time these words were written, such inquiry was severely hampered by Soviet secrecy and societal closure. Today, with the Soviet past more accessible, there is value to be had from revisiting it and drawing appropriate lessons for future reflection.

A LEVEL-OF-ANALYSIS PROBLEM

Between the contrasting Western images from the early 1970s through the mid-1980s that portrayed the Soviet air threat as ten feet tall and three feet short, respectively, it was lost on many protagonists (on both sides of the debate) that they were grappling with a false issue. Each of the opposing images described in detail at the beginning of Chapter Five contained elements of truth as far as it went. Yet each dealt with only a part of the problem. The first gave the VVS too much credit for such nonquantifiable factors as training, tactics, leadership quality, operational prowess, and all the other intangibles excluded from the analysis that, for better or worse, make up the critical link between equipment capability and combat outcomes. The second looked only at the Soviet fighter pilot in isolation and ignored the fact that war is not decided at the 2 v 2 level, but rather by the interaction of countervailing air, land, and naval forces across the board. The cardinal error made by *both* sides was to work one level of aggregation too low in failing to ask how an air force's

---

[1]Raymond L. Garthoff, *Soviet Strategy in the Nuclear Age*, New York, Frederick A. Praeger, 1958, p. xi.

hardware might combine with its operational style and aircrew proficiency to make its influence felt in a campaign context.

Looking at the issue today with the benefit of hindsight and better evidence, we can say with confidence that the edge in that debate belonged to those who saw the Soviet fighter pilot as rigorously trained and technically literate, yet also highly regimented and bound to scripted scenarios heavily dependent on GCI close control, with little room for exercising initiative and virtually no opportunity to develop proficiency at free air-combat maneuvering as Western fighter pilots routinely understood and practiced it.

Complicating the drawing of easy conclusions from that revealed deficiency, however, was an ongoing improvement in Soviet equipment, as attested by the introduction of the MiG-29 and Su-27 into front-line fighter regiments. True enough, improvement in VVS training and tactics proceeded at a snail's pace by comparison. All the same, by the late 1980s the VVS was in genuine intellectual turmoil, and its brightest lights both at senior leadership levels and at the grass roots had come to recognize and admit their shortcomings. Among other things, there was unprecedented discussion of free and engaged roles in aerial combat, as well as debate over the relative merits of single-ship versus team tactics (ultimately decided in favor of the latter, for good reason).

Yet despite these signs of ferment, inertia and continuity for the most part predominated. Soviet fighter aviation remained heavily tied to off-board command and control and reflected deeply ingrained habits that were intrinsic to Soviet culture—not just to VVS culture but to that of the armed forces and society across the board. It was a culture that emphasized the primacy of the collective over the individual. What it produced, and what Russian military professionals now recognize to have been a potentially fatal liability, was an expensively trained fighter pilot with leading-edge equipment who was given little leeway to use it to its fullest capability.

AMBIGUITIES IN THE CHANGING THREAT PICTURE

Does this mean that in an air-to-air Olympiad against Frontal Aviation over the Fulda Gap, the skies of Germany would have been swept clean of Soviet fighters by American and NATO airmen?

Probably. But the question that really matters is: To what ultimate effect? For one thing, the Soviet air threat would not have been the pushover for NATO that the Iraqi Air Force proved to be to the allied coalition in Operation Desert Storm. Like the examples of poor Egyptian and Syrian performance against Israel during a succession of Middle East air battles since 1967, the Soviet-trained Iraqi Air Force bore the heavy imprint of Soviet air-to-air style. But it also represents a highly misleading baseline from which to project how the VVS would have performed in an air war against NATO.

A thoughtful VVS general not long ago admonished me not to equate Russian pilots with Arabs. He had a valid point. Had the Israelis encountered Soviet fighter pilots rather than Syrians in the aerial engagements over Lebanon's Beka'a valley in 1982, there would almost surely have been perceptible differences both in the chemistry of the ensuing combat and in the outcome. To begin with, simply by virtue of their professionalism and upbringing, Soviet pilots would have shown greater air discipline, as well as a purposeful aggressiveness that would have inclined them to stay and fight rather than turn and run when engaged. They most likely would have operated more consistently within recognizable tactical principles. They would have been more knowledgeable about the performance parameters and limitations of their weapons, and therefore better positioned to take full advantage of passing shot opportunities. In the end, however, the outcome would still have been heavily weighted in favor of the Israelis. It would not have been an 85-0 shutout by any means, as the Israeli Air Force accomplished over the Syrians. Nevertheless, Soviet pilots would have ended up on the losing side, because they simply were not trained for the sort of free-form, multiparticipant air combat that ensued once the fights were on.

Had such engagements continued for any length of time, however, Soviet pilots would not have remained hapless losers indefinitely. Notwithstanding their rigidities, the Soviets were (and the Russians remain) capable of purposeful change under stress. Necessity being the mother of invention, they would have licked their wounds and come up with smarter ways, just as they did slowly over the four-year evolution of World War II. The reason such a recovery was never given much credence in the NATO-Warsaw Pact context was that there was little chance of a war lasting long enough (or remaining conventional long enough) to allow such a learning curve time to develop and register its effects.

Even this more circumspect assessment of the VVS's shortcomings is no counsel for complacency, however. Although NATO air-to-air pilots could be assured of going into a fight with a pronounced edge in tactical proficiency over their Soviet opposites, NATO planners and commanders did not enjoy that luxury because they had to worry about a bigger picture. Whatever one might say in hindsight about the individual Soviet pilot and his training inadequacies by Western standards, the VVS fighter force in the aggregate demanded respect. First, it had a definite, if not overwhelming, edge in numbers, which translated into an ability to concentrate force and keep feeding the fight despite high attrition. The VVS further operated within a doctrinal framework that was supremely offensive in orientation. This gave the Soviet side the power of the initiative, plus an advantage in sustaining offensive momentum that naturally accrues to the side with the prerogative of going first. Finally, the Soviet military leadership harbored an attitude toward attrition that did not occasion much concern over the prospect of high loss rates so long as Warsaw Pact ground forces were assured of advancing on schedule at the operational and strategic levels.

THE FALLACY OF MIRROR-IMAGING

This suggests that Western analysts erred whenever they strove to size up the Soviet air threat using *our* measures of effectiveness rather than asking how Soviet planners might assess their own capability. What was needed was an appraisal of the Soviets by their own standards and an explicit recognition that their training activities necessarily took place in a uniquely Soviet context. Soviet commanders may have operated in accordance with a seemingly inflexible operational philosophy. But they were not stupid, and it is highly doubtful that they ever believed that what they were doing was inappropriate to their needs. They knew perfectly well how the USAF and NATO trained, for they could read us like an open book. They also were quite adept at borrowing selectively from Western *technical* practice whenever it suited their needs, as best exemplified by the design features of the MiG-29 and Su-27. Yet despite this, they remained wedded to their own concepts of operations. That they did not elect to emulate Western employment practices with their fourth-generation fighters did not indicate a "slowness to converge." Rather, it revealed a fundamentally different conception of warfare and of the role of air power in it.

Soviet commanders almost surely did not see the training gap in their practices that many in the American fighter community did. The greater likelihood is that they saw their approach to training as *better*, given the way they planned to fight. As one knowledgeable USAF fighter pilot remarked, what might appear to an American observer as "an unimaginative tactic may to the Soviet commander be as sophisticated and advanced as his doctrines, force structure, and mission would dictate. And who is to say that fluid attack and independent maneuvering would work better than regimental control in their battle schemes?"[2]

Simply put, the idea of allowing flight leaders to make autonomous force committal decisions was totally anathema to mainstream Soviet military thought. The General Staff was not only content but also determined to treat its fighter pilots as pawns and to elevate tactical decisionmaking authority to the higher level where, in their view, it properly belonged. American threat analysts would have waited forever for the Soviets to adopt Western operational concepts as demonstrated and refined at Red Flag and in similar large-force exercises around the world. That expectation was a classic case of the fallacy of mirror-imaging in its easy assumption that Soviet development of equipment similar to that of the West would inevitably drive the VVS to adopt similar tactics and concepts of force employment. What was needed, and was all too slow to come among many, was a recognition that the Soviets were marching to their own drummer. Soviet air campaign philosophy, with its heavy combined-arms influence, derived from a unique Soviet military tradition and yielded an image of tactical air power and its role in combined-arms warfare that was dramatically different from philosophies that had evolved in parallel in the West.

Often to the detriment of clear understanding of the dynamics of warfare, it has long been an idiosyncratic trait of the American

---

[2]Colonel Mike Press, "The Human Factor: The United States Versus the Soviet Fighter Pilot," *Air University Review*, November–December 1986, p. 76.

defense-analytic style to carve up war into neatly defined categories, such as air-to-air combat, surface attack, electronic combat, and air-land battle, and then to treat these as though they were hermetically sealed domains of activity with no significant interrelationships or interdependencies. The Soviets, by contrast, saw war as a seamless web. To them, what happened in one category affected activity in all others. In conventional land warfare, air power was regarded as a supporting combat element in a combined-arms approach to force employment. *Everything* the VVS did in air-to-air training had to be viewed within that context to be properly understood. Bluntly stated, only if Soviet air-to-air pilots helped the Soviet front commander accomplish *his* mission of putting a wall of armor on the Rhine River by D-plus-whatever by keeping NATO's ground attack aircraft from slowing up advancing Warsaw Pact tanks and infantry were they performing their assigned function. How they fared in aerial combat itself was completely a side issue. In this regard, Soviet and Warsaw Pact air-to-air pilots were worlds apart from their NATO opposites in mission tasking and expected performance.

With the introduction of longer-range aircraft like the Tu-22M Backfire and Su-24 Fencer, the VVS acquired a range-payload capability that promised to yield something like an independent theater-strategic air offensive option. Nevertheless, there was never an autonomous role for tactical air power in Soviet military thought. Frontal Aviation meant exactly what its designation implied, namely, air power tasked by the front commander to support the latter's operational needs.[3] At the General Staff level, Soviet planners were simply not interested in air-to-air kill ratios as ends in themselves. As Barry Watts has observed, they would have been quite content "to ensure that Pact ground forces will attain their objectives on the desired time lines, even if most of the American F-15 drivers become multiple aces." Their image of the proper use of air power was a carefully crafted offensive air operation, not the putative leverage of

---

[3]This has changed dramatically since the USSR's collapse. As noted in Chapter Three, a recent reorganization of the VVS has brought a newly constituted Frontal Aviation Command out from under its former control by Military District ground forces commanders to report directly to VVS headquarters.

"a handful of pilots trained to a razor's edge."[4] This fact is well captured in the famous cartoon showing two Soviet marshals in Brussels sitting over a bottle of liberated Napoleon brandy celebrating their victory over NATO, at which point one looks to the other and asks: "By the way, comrade, did we win the air war also?"

Even had the VVS lost 80 percent or more of its air-to-air and ground attack assets during a war against NATO, the High Command might have considered that an acceptable buy-in cost if it helped pave the way for a theater victory. This approach to war also posed a non-trivial air discipline problem for defending NATO air-to-air pilots. It was not at all uncommon at the height of the cold war during the early 1980s for young fighter pilots in USAFE to claim with quiet confidence that should war come, they would make ace during the first thirty minutes because of their superior training and air combat prowess. Yet if the price of their making ace on Day One was abandoning their defensive combat air patrol duties and thus enabling VVS strikers to reach their targets deep inside NATO lines, it would have meant making ace for nothing. By contrast, the VVS was *never* in the business of making aces. This point is crucially important for a correct understanding of how air-to-air related to the larger Soviet scheme of war.

THE LIMITS OF INFORMED HINDSIGHT

Had the cold war continued, the USSR would eventually have lost some of its former quantitative advantages as an inevitable result of technological advance. For one thing, the VVS's new fourth-generation fighters represented more complex and costly equipment than it had ever acquired before. That portended a slower production rate and reduced total force size. Although NATO faced a similar problem, the overall trend was nevertheless for a narrowing of the former numerical asymmetry that long favored the Warsaw Pact. A Soviet-American standoff in Europe circa 1995–2000 would not have seen the imbalance of deployed combat aircraft that NATO had suffered in earlier years.

---

[4]Barry D. Watts, "Air Warfare: A Comparison of Soviet and U.S. Views," paper delivered at the Tenth General Working Meeting of the Military Conflict Institute, U.S. Military Academy, West Point, New York, May 4–8, 1987, p. 18.

Furthermore, with the increased complexity of their latest fighters, the Soviets themselves bought into many of the same problems of maintenance and sustainability that the USAF experienced for a time with the F-15, because of its leading-edge engine technology and more sophisticated avionics. To cite but one illustration, as was pointed out in Chapter Two, it was not uncommon for VVS maintenance officers in the late 1980s to complain openly about seemingly intractable problems of radar fault isolation.

Finally, the heightened unit cost and reduced numbers of new Soviet fighters would have made it much more difficult for Soviet planners to continue thinking of their air-to-air assets as attrition fillers. A Soviet air commander in 1995–2000 would most definitely not have enjoyed the luxury of contemplating burning off MiG-29s at the same rate he could have acceptably lost MiG-21s or MiG-23s a decade earlier.

Offsetting these adverse trends, Frontal Aviation by the cold war's end was on the verge of acquiring new capabilities that would have stressed NATO air defenders perceptibly, notably in air-to-air missile range and lethality. It has long been a rule of thumb among fighter pilots that whichever side can get off the first missile shot can control an engagement, at least until the merge. With its new look-down/shoot-down capability, the VVS had reached a point by 1989 where it could deny an F-111 or Tornado crew a confident low-altitude sanctuary against air threats. Finally, with improved infrared missiles offering expanded forward-hemisphere launch envelopes, coupled with a point-and-shoot helmet-mounted sight, even a weak Soviet pilot approaching the merge would have been a threat to respect, since he would have possessed a weapon offering greater shot opportunities than ever before. The fact that he had little by way of basic fighter maneuvers skills would be of only marginal significance for such an engagement scenario, particularly if he had sufficient forward-aspect missile shot discipline prior to reaching the merge.

In all, this juxtaposition of changing strengths and weaknesses leaves us with something of a wash on the question of whether the VVS had vulnerabilities that could have been decisively exploited by NATO. There is little doubt that the USAF would have maintained a commanding edge in air combat maneuvering prowess. On the other hand, the Soviets had compensating advantages in superior numbers, the power of the initiative, an offensive doctrine, greater insensitivity to attrition, and the beginnings of qualitative parity in equipment that would have undermined this edge, at least at the margins. They might also have had more permissive rules of engagement with respect to target identification when it came to beyond-visual-range air-to-air missile employment.

Even today, then, we do not have a conclusive answer to the question of how the VVS would have fared against NATO had it been put to the test of live combat. Now that VVS leaders are freer to discuss such matters openly, and now that it is a legitimate question for cold war history rather than one with intelligence-collection overtones, it might be interesting to draw out the VVS fighter community to ask how *they* saw the issue at the height of the cold war. Such an exchange could make for an informative dialogue between American and Russian fighter pilots and defense analysts. Not long ago, an intelligent and reflective VVS general calmly assured me that "we were never afraid of you." If he meant that, it would be instructive to understand why.

Chapter Eight

RUSSIA'S AIR WAR IN CHECHNYA

The VVS experienced its first trial by fire during the 1994–1995 fighting in Chechnya.[1] Its only combat exposure of comparable note occurred in Afghanistan nearly a decade earlier. Although the Chechen campaign was largely a failed attempt by Russian infantry and armored forces to suppress a local rebellion by military overkill, air power played a prominent part throughout the campaign in providing intermittent support to what would otherwise have been an even more hapless Russian ground contingent.

The drain on resources from being forced into combat at a time of such severe institutional and financial duress was inopportune for the VVS. Nevertheless, the war offered a telling test of the VVS's declared primary role in the post-Soviet era, namely, the projection of air power to conflicted areas along the periphery of the former Soviet Union. Throughout the five years since the allied coalition's success in Operation Desert Storm, Russia's military leaders, notably including those in the VVS, have pointed with respect to the coalition's triumph in the air campaign and have repeatedly held it up as the role model to be followed. They have also claimed, rather optimistically, that despite its manifold problems, the VVS has the needed know-how to deliver an analogous performance.

Because of its topography and weather, to say nothing of the operational challenge it presented, Chechnya was scarcely the clear-cut

[1]This excludes the limited, and largely uncontested, Russian air activity in Tadzhikistan and elsewhere around the southern periphery of Russia following the USSR's collapse.

venue for an air war like the one the allies enjoyed against Iraq. Rather, the VVS's mission entailed backstopping ground troops in putting down an uprising of irregulars more than it did frontally engaging organized formations and attaining well-defined objectives on a battlefield. There were few fixed targets of major military significance in Chechnya, and no front lines whatsoever. This rendered the situation more like what NATO recently faced in Bosnia than what the allied coalition had to contend with in Desert Storm. Nevertheless, Chechnya provided a realistic and relatively low-risk laboratory for testing the new Russian strategy, as well as the VVS's operational capabilities under live-fire conditions. Needless to say, the returns were mixed.

The VVS's airlift arm carried its share of the burden in Chechnya commendably, despite severe serviceability problems and its substantial loss of assets to Ukraine and Kazakhstan (almost half its 500 Il-76 jet transports) following the USSR's breakup. VVS combat aircraft also performed well in unopposed ground attacks against unsheltered Chechen aircraft during the war's preliminaries. However, as the initial ground campaign unfolded and stresses mounted owing to weather complications and the demand for high-accuracy bombing in the face of effective low-altitude Chechen air defenses, degradations in VVS performance displayed on repeated occasions clearly attested to the deprivation it continues to suffer in curtailed training because of lack of money.

This was the first Russian military venture to be conducted under the full glare of international press attention. Never before had outsiders been able to monitor Russian military operations so closely as in the case of Chechnya. Not surprisingly, the increased openness of post-Soviet Russia made for a recurring approach-avoidance conflict among military authorities. On the one hand, the beleaguered and badly underfunded Ministry of Defense, clearly intent on making the best of this opportunity to "show its stuff," sought to use the war as an occasion to cast its strengths in the most favorable light, as well as to win sympathy for the privations it had been forced to suffer since the collapse of Soviet communism.

On the other hand, the High Command continued to carry a lot of Soviet baggage in its day-to-day operating routines. So burdened, it was plainly discomfited at having its every action observed and criticized by the media. Military leaders repeatedly fell back on secrecy and security arguments to evade pointed questions. They also sought, sometimes aggressively, to hinder the work of reporters.

For the most part, General Deinekin has spoken candidly about the VVS's combat performance and problems in Chechnya. Admittedly, he has remained mute on such crucial matters as the tempo and level of intensity of air operations, the total number and types of aircraft employed, the number of combat and combat support sorties flown, and weapons delivery modes used in ground-attack operations. Morever, he has provided little by way of a detailed accounting of weapons effectiveness and operational results. Nevertheless, he has disclosed enough to allow us to see at least the big picture, something that would have been difficult at best in the case of any comparable Soviet experience in years past. Based on his remarks and various statements of other Russian military leaders, not limited solely to those in blue uniform, this chapter considers the role played by Russian air power in the Chechen war and what the performance of the VVS—and of the Russian military more generally—tells us about Russia's changing approach toward the use of force and its near-term military potential.[2]

HIGHLIGHTS OF THE AIR CAMPAIGN

Chechnya is a mountainous enclave in Transcaucasia approximately the size of the state of Connecticut, with a population of 1.3 million people, including a multitude of truculent and feuding ethnic clans. It is legally a part of the Russian Federation, having been annexed by Russia during tsarist times. For reasons too complex to review here, it constituted a time bomb that was bound to go off in Moscow's face sooner or later in the wake of the USSR's demise.

To summarize the origins of the conflict, separatists in the Chechen-Ingush region saw a ripe opportunity for secession building as the

---

[2]A competent Russian treatment of the Chechen war that draws on many of the sources used in this chapter was published in June 1995 under the lead authorship of Dr. N. N. Novichkov, deputy director of the ITAR-TASS Agency for Scientific and Technical Information. For a translation of the chapter on air operations, see *Frontal and Army Aviation in the Chechen Conflict*, Conflict Studies Research Center, Royal Military Academy Sandhurst, Camberley, England, June 1995.

USSR careered toward collapse in 1991. On August 21, the third and final day of the failed Soviet coup attempt, Chechnya commenced its disengagement and declared independence from the Russian Federation the following September 6. Two months after the abortive coup, Major General Dzhokar Dudayev, a former VVS bomber pilot and air division commander, was elected president.[3] He reaffirmed Chechnya's independence and promptly established an iron rule over what Moscow defense correspondent Pavel Felgengauer has aptly labeled a "strange buccaneer republic."[4]

Moscow first ignored the problem, then tried twice to topple Dudayev through clandestine operations. The first attempt, which bore heavy marks of involvement by the Federal Counterintelligence Service (FSK), strove to exploit opposition to Dudayev among the Chechen population. That ill-fated effort unraveled in late 1994 when Chechen rebels successfully countered it. The second attempt, on November 26–27, failed catastrophically, with Dudayev claiming 67 Russian tanks destroyed. A Russian airborne spokesman later admitted that the attack was doomed from the outset because "without infantry cover, it was really senseless to bring tanks into the city."[5] After the embarrassment of this second failed attempt,

---

[3]General Deinekin, under whom Dudayev served while Deinekin was LRA commander, had temperate words for Dudayev during the early days of the confrontation: "Dudayev was an intelligent commander, a highly-qualified pilot—we never put pilots with a poor reputation in charge. . . . Dudayev was quite a good pilot, a good commander, he was known for his concern for people, he was very efficient and dependable." Interview by Vladislav Listyev on Ostankino television First Channel, December 14, 1995. By the end of March, Deinekin merely noted that Dudayev had commanded an air division "which was no worse than others" and that "the general himself did not shine with any outstanding talent, although he performed meticulous service." Deinekin added that the Muslim world was well aware of Dudayev's bomber division's combat operations against the mujaheddin in Afghanistan and that for this reason, Dudayev "had to change his image." Interview by Yury Dmitriyev and Nikolai Kishkin, "Air Force Commander in Chief Petr Deinekin: 'I Am Prepared to Account for Every Aerial Bomb . . .,'" *Trud*, March 2, 1995.

[4]Pavel Felgengauer, "The Chechen Campaign," paper presented at the Third Annual Conference on Russian Defense Decisionmaking, U.S. Naval Postgraduate School, Monterey, California, November 7–8, 1995, p. 6.

[5]Quoted in Timothy L. Thomas, "The Russian Armed Forces Confront Chechnya: II. Military Activities, 11–31 December 1994," *Journal of Slavic Military Studies*, June 1995, p. 268.

President Yeltsin decided to move directly by committing Russian forces to a massive intervention on the ground.[6]

Chechnya presented Russia with a nettlesome difficulty. Indeed, Moscow's "Chechen problem" had roots running well back to pre-Soviet times. Neither the tsars nor the communists had fully succeeded in subjugating the fiercely independent Muslims who populated Chechnya and who had sustained a simmering hatred for Russians ever since their forced assimilation. During World War II, Stalin deported the Chechens wholesale to Kazakhstan out of fear that they would otherwise collaborate with the Nazis. Khrushchev finally permitted them to return home in the mid-1950s, whereupon they discovered that everything they had left behind had been taken over by Russians. This triggered a vicious underground ethnic campaign against the despised Russians, who routinely fell under indiscriminate knife killings at the hands of enraged and embittered Chechens. Russian vigilante groups soon formed up to return the compliment in kind, with the result that Soviet troops had to be called in by 1959 to restore order.

Nevertheless, the violence continued to simmer at a lower temperature, and Chechnya remained a dangerous place for Russians. Lieutenant Viktor Belenko, the MiG-25 pilot who defected with his aircraft to Japan in 1976, received basic flight instruction at the nearby Armavir flight school, which maintained an auxiliary training field just outside the Chechen capital of Grozny. Belenko recalled that the KGB officer who had given his class a local-area orientation after they had first reported to Grozny made a point of warning that "most of all, you must guard yourself against the Chechens. The Chechens use knives wantonly, and under stress they will butcher

---

[6]This immediately prompted speculation in Moscow that the defense ministry had not planned a tactically sound operation to settle the Chechen problem but instead was suckered into a decision by cabal under pressure from Yeltsin's closest advisers. See, for example, Nikolai Vishnevskii, "Grozny Offered a Dialogue," *Nezavisimaia gazeta*, December 17, 1994. This report called the failed clandestine assault on Grozny on November 26 "manifestly amateurish in nature" and an operation conducted by forces other than regular military, notably the former KGB and the Ministry of the Interior. Many accounts saw the defense ministry blindsided by Yeltsin's closest cronies. For a detailed review of events leading up to the invasion and the politics behind them, see Timothy L. Thomas, "The Russian Armed Forces Confront Chechnya: I. Military-Political Aspects, 11–31 December 1994," *Journal of Slavic Military Studies*, June 1995, pp. 233–256.

you."[7] Belenko further remarked that many Chechens had been reared from birth to believe that one could never attain full manhood without first killing at least one Russian.

True to this tradition, the government presided over by Dudayev was illegitimate and irresponsible in equal measure. To say the least, it offered safe haven to an assortment of undesirables who populated Russia's criminal underworld, and it was no more accountable than the Yeltsin government in failing to negotiate a peace settlement, or at least a modus vivendi, throughout the three years before Russia's armed intervention. In the spring of 1993, President Dudayev dissolved Chechnya's parliament and proceeded to rule as a warlord. Shortly thereafter, dissenters clashed with Dudayev's presidential guard, suffering nearly 50 killed in the process. Dudayev went out of his way on repeated occasion to irritate Moscow. Chechen bandits seized hostages for ransom in several cities in southern Russia, and the capital city of Grozny earned a deserved reputation as a hotbed of criminal gangs, as well as, by one account, "the destination of choice for anyone hijacking a Russian airliner."[8] Moscow's tolerance of such behavior was bound to wear thin. By November 1994, the forbearance of the Yeltsin government finally broke.

The Operational Setting

Chechnya's force structure, such as it was, consisted of arms and equipment left behind by the departing Soviet military following the breakup of the USSR. Spoils accruing to the Chechen ground forces included 42 Soviet tanks (a mix of T-62Ms and T-72s), 66 armored combat vehicles, 18 Grad multiple rocket launchers, 30 122mm towed howitzers, and 523 RPG-7 antitank grenade launchers.

Chechnya's air defenses included four mobile ZSU-23/4 radar and optically tracked antiaircraft guns, six ZU-23 and DShK optically sighted machine guns, portable grenade launchers, and small arms mounted on trucks and passenger cars. General Deinekin said the

[7]Quoted in John Barron, *MiG Pilot: The Final Escape of Lieutenant Belenko,* New York, McGraw-Hill Book Company, 1980, p. 68.

[8]S. Frederick Starr, "Chechnya: The U.S. Interest," *Wall Street Journal,* December 22, 1994.

Chechen irregulars also possessed "several thousand" shoulder-fired infrared SAMs.[9] The latter included the SA-7/14 Strela/Gremlin and possibly also the SA-16 Igla/Gimlet. Neighboring Azerbaijan was said to have provided Dudayev with additional infrared SAMs.[10]

Chechnya had only a limited air capability consisting of 152 Czech-built L-39 jet trainers, 94 older L-29s, and several MiG-15 and MiG-17 first-generation Soviet jet fighters inherited from the Armavir flight school.[11] Some Chechen pilots reportedly received continuation training in Azerbaijan after the USSR's collapse. Chechnya was said to have "several dozen" fully trained military pilots, along with some mercenary pilots hired from the former Soviet republics. Chechen L-39s were believed to have taken part earlier in the Georgian-Abkhazian war by bombing Georgian positions, with notable effect. So their combat potential was more than hypothetical.

Air operations in the Chechen war played themselves out in three phases: (1) preparatory moves, (2) the gaining of air control, and (3) sustained air support to ground operations. According to one account, only "several dozen" combat aircraft took part in these operations.[12] Considering another report that as many as 26 VVS aircraft sustained battle damage during the course of the war, however, this number is probably low by a considerable margin.

Preparatory Moves

Russian air operations were conducted by units from the VVS's Military Transport Aviation (VTA), Frontal Aviation Command (KFA), and Long-Range Aviation (LRA), as well as by assets from the separate Russian Air Defense Forces (VPVO) and army rotary-wing aviation. The latter included both attack and transport helicopters. Russian naval aviation did not participate in the war.

There were early reports of undeclared VVS participation in air strikes against Chechnya both before and during the second incur-

---

[9]ITAR-TASS, Moscow, January 11, 1995.

[10]Interfax, Moscow, October 10, 1995.

[11]Felgengauer, "The Chechen Campaign," p. 5.

[12]Vremya television program, Moscow, June 28, 1995.

sion on November 26, including alleged attacks on a tank unit at Shali and other missions against Braginskii, Terskii, Katayama, and the Grozny North airfield.[13] The first formal air involvement occurred later that month in response to claimed intelligence reports that a peaceful settlement was out of the question.

VVS combat aircraft used in the war were drawn mainly from units reconstituted from disbanded Soviet ground attack regiments formerly based in Eastern Europe. After the Warsaw Pact's dissolution, these reformed units had relocated to bare bases in the North Caucasus Military District, which offered little infrastructure and only half the needed fuel, rations, ammunition, and spare parts.

Su-24MR reconnaissance jets conducted detailed photography of the three Chechen airfields of Kalinovskaia, Khankala, and Grozny North. They also collected target information on potential military objectives in Grozny and its suburbs and in other areas. According to VVS accounts, they produced "conclusive evidence" that Dudayev was gearing up for combat. Fortified areas were being erected and L-29 and L-39 jet trainers were being prepared for possible use.[14] Dudayev's forces had been detected by VPVO to be preparing highways and road segments as alternate runways to accommodate flight operations. At least some of the L-39s were configured with wing stations for carrying 100-kg bombs and rocket pods.[15] These aircraft could have been used against Russian troops, as well as against such lucrative targets as nuclear reactors, chemical plants, and weapons storage dumps. By the VVS's admission, however, they lacked sufficient range to reach Moscow, particularly if loaded with munitions.[16]

---

[13]See Charles Blandy, "The Battle for Grozny," *Jane's Intelligence Review*, Volume 7, No. 2, February 1995, pp. 53–56.

[14]Lieutenant Colonel V. Beltsov, "Air Operations in Chechnya," *Vestnik vozdushnogo flota*, 1995.

[15]Asked where Chechnya got its aircraft, General Deinekin answered emphatically that "Dudayev did not get from the [Russian] air force a single screw, landing gear, or cotter pin, much less a combat aircraft. Chechnya appropriated DOSAAF aircraft and the air defense and Aeroflot flying schools." Interview in *Trud*, March 2, 1995.

[16]See Aleksandr Khokhlov, "Will Dzhokar Dudayev Bomb Moscow?" *Komsomolskaia pravda*, January 17, 1995.

The invasion by Russian ground units was preceded by a VTA airlift into Mozdok, just to the northwest of the secessionist republic, on November 30. Poor weather hampered the airlift. Nevertheless, VTA played a key role in the buildup, which totaled some 40,000 troops at the height of Russia's involvement in the fighting. It was reported on December 1 that 38 VTA An-12 transports with troops and equipment had flown into Vladikavkaz airport in adjacent North Ossetia. A report the following day noted that airborne troops from the Tula airborne division had been delivered to Mozdok. Equipment airlifted into the war zone included tanks, multiple rocket launchers, self-propelled howitzers, bridge-laying components, and support vehicles. This hardware came from stocks of the North Caucasus Military District. The troops were a mix of officers and conscripts from the Ministry of Internal Affairs (MVD), mobile-force light infantry brigades, and airborne assault forces. Two squadrons each of Mi-24 Hind attack helicopters and Mi-8/17 Hip/Hip-H medium transport helicopters were also deployed to the theater to support impending combat operations. These were supplemented by Mi-26 Halo heavy-lift helicopters and an Mi-9-equipped command and control center.

VPVO played an active role by maintaining round-the-clock A-50 Mainstay AWACS tracks on all sides of Chechnya to monitor air traffic coming in or out. On November 30, the VPVO's commander in chief, Colonel General Prudnikov, indicated that his command had been under orders since the preceding August to "close" Chechen airspace so as to prevent the influx of military assets of any kind to Dudayev. Since then, he said, no aircraft had landed in Chechnya or departed without his personal authorization. VPVO also kept two to six MiG-31s or Su-27s on constant combat air patrol (CAP) to intercept any aircraft that might resupply Chechnya or threaten Russian troops on the ground. These were the only fighters employed in the war. Since there was no air-to-air threat, the MiG-29 was not a player.

By the end of January, VPVO interceptors had logged some 1500 hours on combat air patrol to blockade Chechnya's borders from external resupply. The chief of the VPVO headquarters staff, Colonel General Sinitsyn, reported that the A-50 AWACS covered "virtually the entire region" and was supplemented by low-level gap-filler coverage provided by individual VPVO radar platoons and companies.[17]

A serious bottleneck during this initial phase was the overburdened Mozdok airfield, where most Russian military aircraft were concentrated. By one account, air traffic control personnel suffered a "colossal load," and aircrews recovering from combat missions "were literally forced to maneuver around to land their aircraft safely on the runway." VTA delivered most military personnel first into Mozdok and Vladikavkaz airfields, and later directly into Grozny North once the latter was secured by Russian ground units in January.

The Gaining of Air Control

The second phase of the air war overlapped the first and entailed taking out Dudayev's limited air force in a three-day airfield attack operation on November 28–30. This was not an imposing task, since the aircraft were unrevetted and Dudayev presented no counterair threat. By one account, only six Su-25s out of VVS bases in the North Caucasus Military District were used in bombing and rocket attacks on the three Chechen airfields of Khankala, Kalinovskaia, and Grozny North.[18] These attacks destroyed or neutralized all 266 Chechen aircraft, including one Tu-154, six Tu-134s, and three helicopters in addition to the L-29s and L-39s. Little collateral damage was caused in the process to runways and taxiways, airport installations, and radio and lighting equipment. Immediately after the destruction of his L-29s and L-39s by the VVS, Dudayev wired defiant congratulations to General Deinekin, stating: "I congratulate you and the Russian VVS on another victory in achieving air superiority over the Chechen Republic. Will see you on the ground."[19]

---

[17]Aleksandr Ivanov, "Dudayev Has No 'Air Bridges,'" *Krasnaia zvezda*, March 21, 1995. Sinitsyn added that these VPVO radar units were as deprived of creature comforts as any other Russians deployed, noting that they had no tents for heating, bathing and laundry facilities, cooks, or even an adequate change of warm underclothing.

[18]Simon Elliott and Alexander Velovich, "Backing Down: Russian Air Attacks on Chechnya Seem to Have Failed," *Flight International*, January 11-17, 1995, p. 23.

[19]Interview with General Petr S. Deinekin, "The Air Force Between the Sky and . . . the Scandals," *Argumenty i fakty*, December 1994, p. 8.

Air Support to Ground Operations

The third phase of the air war began with the advance of Russian ground troops toward Grozny on December 11. Concurrently, the weather took a turn for the worse, confronting VVS aircrews with ground fog, blowing snow, severe icing, and a heavy cloud buildup, with a low ceiling and tops above 15,000 ft. This made both high- and low-angle manual bombing impossible and also precluded any resort to electro-optical or laser-guided weapons. Instead, the VVS was forced to employ Su-24s in day and night level bomb releases from medium altitude (15,000–20,000 ft) against radar offset aim points, or in inertial bombing against geographic coordinates, through heavy cloud cover. The gross inaccuracy of these weapons deliveries resulted in many Russian troop losses to friendly fire.

Russian defense minister Pavel Grachev, who organized and commanded the operation, made slow and indecisive use of his air assets, just as he did with his forces on the ground. There was a report on December 14 that five Su-25s had carried out a late-afternoon rocket and strafing attack against targets in the center of Grozny.[20] Not until December 19, however, was the Grozny television tower brought down. News reports on December 22 confirmed that the VVS was bombing Grozny with Su-24s and Su-25s operating out of the military airfields at Yeisk and Budennovsk. By Christmas Day, 80 percent of Chechnya lacked electricity, and gas supplies to half the country had been severed as a result of the air attacks.

Early in the war, reported bombing inaccuracies underscored the pilot proficiency shortfall the VVS had been forced to endure as a consequence of four years of deprived funding for training. Most VVS aircrews who participated in the initial attacks had not flown more than 30 hours the preceding year. Few were night-current or maintained any precision weapons delivery proficiency, if indeed they ever had any. As a result, General Deinekin was forced to assemble a "tiger team" from among his most experienced weapons instructors and test pilots to send to the war zone. Only then did battlefield air interdiction operations begin to show positive results. For a time, Deinekin experimented with a "blue-gold" aircrew arrange-

---

[20]ITAR-TASS, Moscow, December 14, 1994.

ment, whereby the crew ratio in the theater would be doubled and aircraft could be turned more quickly with fresh pilots.[21]

On December 22, four Su-24s used FAB-500 1000-lb general purpose bombs to attack the highway between Staraya Sunzha and Verkat-Yurt northeast of Grozny to preclude rebel gunmen from moving into the city. Once the weather broke on December 29-30, VVS pilots used electro-optical and laser-guided weapons against Chechen command posts, as well as to drop bridges over the Argun River 10 km east of the city to prevent Dudayev from bringing reserves into Grozny. Their destruction was later confirmed by reconnaissance overflights tasked to provide battle-damage assessment.

General Deinekin admitted that the VVS had "regrettably" used cluster bombs against rebel force concentrations, but denied that it had employed flechette weapons banned by international convention.[22] The VVS used parachute-retarded flares during night operations to provide illumination for visual bombing whenever the weather permitted. General Deinekin also noted that in some cases, the VVS flew intentional low supersonic passes over Grozny, laying down sonic booms to simulate bomb explosions to intimidate the rebels.

Later, on January 17–18, seven Su-25s carried out a high-accuracy attack against the presidential palace, using rockets and concrete-penetrating BetAB 3000-lb unguided bombs. Two of these bombs penetrated the palace from top to bottom. Another five took out an underground tunnel and a command post buried deep beneath an adjacent building.[23] Also, an arms dump on the northern outskirts of

---

[21]See Yuliya Kalinina, "Operation 'New Year': Grozny Was Bombed by Diesel Engine Mechanics in Training to Be Pontoon Bridge Builders," *Moskovskii komsomolets,* January 6, 1995.

[22]Interview with Colonel General Petr S. Deinekin, "Flying in Your Dreams and in Reality," *Rossiiskiye vesti,* August 17, 1995.

[23]To dramatize the alleged effectiveness of these strikes, the Russian Ministry of Defense released the following intercepted radio transmissions from rebel forces in the Chechen command post:

"Cyclone to Panther 1. We're being bombed. They're blowing holes in the building right down to the cellar."

"Get all the leaders together in the large hall."

Arshty was destroyed by Su-25s. The attack on the presidential palace was led by an experienced pilot, most likely detached from the VVS's weapons training center at Lipetsk.

The first VTA transport deployed into Grozny North on January 18, after which construction personnel from army aviation radiotechnical support battalions installed a control tower, radio and landing aids, a navigation beacon, and runway and taxiway lights. They also brought in fire trucks and washdown vehicles. Since then, the airfield has continued to operate at full capacity.

On January 25, eight Su-25s attacked Chechen underground ammunition dumps located in four former Soviet ICBM silos. These were destroyed, along with an underground battalion command post, ventilation intakes and vents, mobile antennas, an installation and testing building, and adjacent trailers piled high with ammunition.

Mobile air defenses operated by the Chechen resistance were close-controlled by radio. They shifted position constantly, further impeding their detection and destruction by Russian forces. Dudayev shamelessly positioned ZSU-23/4 and infrared SAM defenses in the midst of densely populated residential areas. Their effect was "pretty fierce," according to General Deinekin. Portable SAMs "did not see high use," although several were fired against both fixed-wing aircraft and attack helicopters. The VVS used flares on occasion to counter infrared SAMs.

The first VVS aircraft downed was a Su-25 on February 4. During an attack on a rebel strong point 2 km south of Chechen-Aul, a two-ship

---

"They're using direct fire against the command post."

"We need to withdraw our forces to the other side of Sunzha. Otherwise they'll bury us."

"The second line of defense will be at the Minutka intersection. There are many dead and wounded in the palace. There isn't time to deal with them. We need to get out ourselves. If we don't manage it now, we'll wait until dark and leave."

This was followed two hours later by another transmission:

"Panther 3 to Cyclone. After today's strike, everyone is very badly shaken, in shock. The strikes were very powerful and precise."

Quoted in Beltsov, *Vestnik vozdushnogo flota.*

element of Su-25s was working a bridge on the Argun River when a ZSU-23/4 opened up on both, bringing one down and killing its pilot, Major Nikolai Bairov.[24]

Russia's rotary-wing aviation likewise received a renewed baptism of fire in Chechnya. It had experienced only limited previous combat exposure in Abkhazia, South Ossetia, Transdniestria, and Tadzhikistan, all related "hot spots" that had sprung up in the wake of the USSR's collapse.

The Afghan experience offered a solid foundation upon which to improvise helicopter operations and tactics. Mi-24 Hinds utilized target approaches roughly comparable to those battle-tested in Afghanistan. Techniques used in Chechnya included nap-of-the-earth (NOE) operations, approaching an objective from alternating directions, randomly jinking before a final attack run, executing an NOE egress with jinkouts, and heavy reliance on mutual fire support, electronic countermeasures, and flares. Said one expert: "Life forced constant corrections to our operating tactics." More often than not, Russian intelligence on Dudayev's limited air defense dispositions was poor to nonexistent. This forced helicopter crews to operate outside the lethal envelopes of rebel air defense weapons.

In a related problem, the high density of enemy defenses in certain areas made it impossible for helicopters to use antitank guided missiles (ATGMs) against hardened structures. Instead, attack helicopter crews were forced to resort to S-24 high-velocity unguided rockets. The effective slant range of these rockets was only 3000–5000 ft, which put the helicopter inside the engagement parameters of enemy defenses and dictated a never-before tested tactic of launching rockets out of a rapid pitchup and pushover. One concern was that the helicopter's engine might fail during the maneuver as a result of air starvation due to the ingestion of rocket exhaust fumes.

Ground forces aviation commanders were not ready to risk taking hits from rooftop snipers. Their commander in chief, Colonel

---

[24]Reports from Moscow indicated that the pilots of three VVS aircraft ejected successfully but were executed by Dudayev's forces. They also said that the Chechen terrorist Shamil Basayev had shot Russian pilots (none of whom had fought in Chechnya) during his bloody rampage in Budennovsk. Vremya television report, June 28, 1995.

General Vitaly Pavlov, stated that it was formal doctrine of his command that "urban combat is not suited to helicopters." The latter, he said, are most effective against open-country targets, using unguided rockets and the AT-6 Shturm guided missile.[25] (According to Pavlov, Russian attack helicopters did not drop gravity bombs.)

The Chechen resistance made widespread and effective use of ambush tactics, concealing their presence and starting to shoot from multiple directions once a helicopter entered their zone of fire. Typically a helicopter returning to base after sustaining battle damage would indicate hits from multiple directions and multiple weapons. An experienced pilot graphically recalled one such ambush involving a two-ship element of Mi-24s in three successive combat sorties on a mission near Gudermes. In each case, the helicopters made preliminary landings en route to update their target information. The third time this the now-predictable element entered the combat zone, it took intensive fire from three directions. One Mi-24 was downed, although the crew was safely extracted, and the other sustained damage. Rued the pilot afterward: "This is how stereotype is punished in war."[26]

Dudayev's tactical intelligence elicited grudging Russian respect. Russian attack aircraft and helicopter radio call signs were changed daily. Nevertheless, in the account of one Russian participant, "one had a feeling that they [the Chechen irregulars] knew a great deal."[27] Rebel forces made a special effort to hunt down Russian forward air controllers (*avianavodchiki*), of which some forty had been attached to the ground forces. In one example cited, no sooner had a FAC gone on the air near Chechen-Aul than massive shelling commenced on his position. The rebels did good work triangulating his location until a Russian motorized infantry unit finally pinpointed and seized the offending direction-finding equipment. FACs in Chechnya were said to have performed better than in the earlier case of Afghanistan,

---

[25]Quoted in Paul Beaver, "Army Aviation in Chechnya," *Jane's Defense Weekly*, June 10, 1995, p. 79.

[26]Colonel Anatoly Surtsukov and Lieutenant Colonel Sergei Prokopenko, "A Shooting Sky: How Many Years Now Has It Tested the Mettle of Ground Forces' Helicopter Pilots?" *Krasnaia zvezda*, July 18, 1995.

[27]Ibid.

although they were hampered by obsolete communications and navigation equipment.[28]

After Dudayev's forces escaped Grozny, the VVS unleashed daily air attacks over a two-month period against the outlying road net and associated villages harboring enemy units to deplete resistance assets and secure Russia's position on the ground. From early March through June, the weather stayed generally cooperative, making possible round-the-clock battlefield air interdiction, as well as photoreconnaissance, battle damage assessment, and attacks with precision munitions, including the AS-12/14 missiles and KAB 1000-lb laser-guided bombs.

President Yeltsin tried to put the best face on a grim situation in mid-January by peremptorily declaring the war won and turning occupation duties over to the troops of the Ministry of Internal Affairs (MVD), while at the same time firing the war's most vocal critics in the senior military ranks.[29] In a transparent bid to invoke the image of the Red Army's triumphant hoisting of the Soviet flag over the Reichstag in Berlin in 1945, the Yeltsin government ostentatiously announced the raising of the Russian Federation flag over the gutted presidential palace in Grozny on January 19. This, in turn, prompted an effort by nationalist elements in the Russian parliament to propose awarding the commander of the 276th Motorized Infantry Regiment the title of Hero of Russia.[30] The proposition was voted down.

A fleeting truce from May 1 to 12 gave way to renewed fighting and a resurgence of VVS air attacks. By this time, Dudayev's forces had taken refuge in the mountains and the Yeltsin government found itself embracing a tarbaby in Chechnya akin to England's in northern Ireland. General Deinekin predicted that rebel operations on the

---

[28]Among the best of the VVS's forward air controllers in Afghanistan were former pilots or navigators who had been removed from flight status for medical or other reasons. See Colonel I. Alpatov, "Forward Air Controller: The Experience of Combat Operations in Afghanistan," *Aviatsiia i kosmonavtika*, No. 5, May 1990, pp. 18–19.

[29]See Michael Specter, "Yeltsin Declares Chechen War Over, and Ousts Four Critics in Military," *New York Times*, January 20, 1995, and Fred Hiatt, "Moscow Debates Intensify Despite Claims of Victory," *Washington Post*, January 21, 1995.

[30]See Steven Erlanger, "High Price of a 'Victory,'" *New York Times*, January 22, 1995.

ground would resume in mid-autumn of 1995, since inclement weather would again work to restrict air operations.[31] He was correct. He further predicted that in the spring, when the trees are dense with foliage, aerial reconnaissance would be less effective against rebel strongholds in the mountains. More somberly, Grachev conceded that the situation could devolve into a partisan war, with rebel hit teams operating clandestinely at night all over Chechnya trying to penetrate Russian positions. He said that such a slow-motion, bleeding war of attrition could continue "for a lengthy period of time. For months, if not years."[32]

A turning point in the war came on June 14, 1995, when a Chechen guerrilla team led by Shamil Basayev carried out a successful out-of-area raid in the Russian town of Budennovsk. This terrorist operation, which saw over 1000 civilians held hostage in a city hospital for six days, left at least 123 dead and triggered two abortive counterattacks by Russian security forces. The event thoroughly swung Russian popular opinion against the war and prompted an eventual cease-fire once Prime Minister Viktor Chernomyrdin secured a bloodless end to the hostage crisis by promising negotiations in return for a release of the captives and safe passage for the Chechen guerrillas.

Today, the heaviest fighting is over and a Russian-appointed administrator has been installed in Grozny. There remains no end in sight, however, to Russia's military involvement in Chechnya. In a renewed outbreak of mass violence, Russian forces sealed off Gudermes, Chechnya's second largest city, on December 14, 1995, and shelled it indiscriminately for eleven days, killing an estimated 600 people (half noncombatant civilians) after Chechen rebels seized the city commandant's headquarters and held 130 MVD troops captive.[33] Although the battle resulted in a Chechen retreat, it signaled an end to the shaky truce that had been in effect since July and left Moscow, as before, with no more than a Carthaginian peace in

---

[31]Information Agency *Ekho Moskvy*, August 19, 1995.

[32]Interview on Ostankino television, March 31, 1995.

[33]See Richard Boudreaux, "Russians Seal Off City, Shell Chechen Civilians," *Los Angeles Times*, December 25, 1995, and "Death Toll Reported at 600 from Fighting in Chechnya," *New York Times*, December 26, 1995.

Chechnya. It was followed in early January 1996 by the siege of Pervomaiskoye by Russian ground and air units, which ended in an ignominious escape by most of the Chechen rebels who had sought to stage a replay of Budennovsk after several weeks of confrontation. Commenting on the general haplessness of the Yeltsin government's strategy toward Chechnya, one Russian reporter concluded presciently after the failure of the initial assault on Grozny in December 1994 that "Foreign Minister Kozyrev has a powerful new argument he can easily use for opposing admission of the former socialist countries into NATO. . . . There is no need to be afraid of us. We cannot do anything anyway. And Chechnya is the best confirmation of this."[34]

Results and Costs

By the end of January 1995, reported equipment losses to enemy fire by Russia's ground forces numbered more than 100 tanks and twice as many infantry fighting vehicles and armored personnel carriers.[35] The Russians lost as many men during the first month of the Chechnya operation as the Soviets did during the first six months of the far larger Afghan war. For their part, during the first three months of combat operations after the attainment of air control, Russian aircrews destroyed three Chechen helicopters, 20 tanks, 25 armored personnel carriers, 130 cars, seven bridges, six self-propelled antiaircraft weapons (including SA-9s and SA-13s), and an artillery battery.[36]

The price of Moscow's miscalculation of what it was getting itself into in Chechnya was dear. In the harsh judgment of the deputy executive director of the Russian Council on Foreign and Defense Policy, the debacle "exposed to the whole world (including not only Russia's friends) that the Russian leadership and Russian armed

[34]Kalinina, *Moskovskii komsomolets*, January 6, 1995.

[35]"The Total in Writing: Is Price No Object?" *Obshchaia gazeta*, No. 3, January 19, 1995, p. 5.

[36]David A. Fulghum, "Chechnya Cripples Russian Aviation," *Aviation Week and Space Technology*, August 7, 1995, pp. 20–21.

forces cannot resolve militarily even a limited conflict."[37] By the official count of the MVD, 1867 Russian troops were killed in action and 6481 wounded, with 36 still being held prisoner by the Chechen rebels. Contesting these figures, the Chechen government head, Salambek Khadzhiyev, has claimed over 4000 Russian fatalities, adding that 6000 Chechen civilians died in the fighting. He said that the war had produced over 400,000 civilian refugees.[38] Since truth is the first casualty in war, there is no telling where the correct numbers lie among these conflicting claims. There is no denying, however, that Moscow's sacking of Grozny produced an all-around human tragedy by any measure. Western reporting at the end of 1995 indicated that the Chechen war had taken at least 20,000 lives altogether.[39] Today, that number is placed at closer to 35,000.

The war was extremely unpopular within the military. A report in April by the head of the defense ministry's Main Personnel Directorate indicated that 557 officers who had refused to fight had been dismissed from service, and some served with criminal charges.[40] In a further aggravation of the general discontent within the ranks, the heavy tapping of Russian war reserves that was required to support the Chechnya operation forced many military districts to reach into their emergency rations to feed their personnel.

The war claimed the lives of four VVS airmen through the downing of two Su-25s and a Su-24 by Dudayev's forces. At least 26 VVS aircraft sustained battle damage. Ten Russian helicopters were shot down and two crews were summarily executed by Chechen rebels. Chechen air defenses, which General Deinekin credited as having been "very effective," produced these results without the support of a single surveillance radar. There were reports as well of American Stinger shoulder-fired infrared surface-to-air-missiles (SAMs) in Chechen hands. This seems unlikely, considering that weapon's

---

[37]Alexander A. Belkin, "The War in Chechnya: Impact on Civil-Military Relations in Russia," paper presented at the Third Annual Conference on Russian Defense Decisionmaking, U.S. Naval Postgraduate School, Monterey, California, November 7–8, 1995, p. 17.

[38]"Chechnya: They Didn't Stint on the Cost," *Trud*, August 1, 1995.

[39]Stephanie Simon, "Russians Offer Amnesty to Chechens," *Los Angeles Times*, December 3, 1995.

[40]Report by Aleksandr Gerasimov on Moscow NTV, April 7, 1995.

proven effectiveness against the Soviets in Afghanistan. Had the Chechens possessed Stingers, air losses by the VVS and army aviation could have been considerably higher.[41]

Chechen agents have been aggressively seeking to gather up the names of Russian pilots who flew combat missions against Grozny in connection with what appears to be an assassination plan for revenge. General Deinekin indicated that bands of Chechens had been observed reconnoitering VVS bases, but that Russia's security services were taking "appropriate countermeasures."[42] That assurance offered cold comfort to the Russian airmen affected by the implied assassination threat. The Chechens are a patient people with a deep determination to settle old scores. Any Russian pilots involved in the war who are even remotely aware of this may have a hard time sleeping soundly for the rest of their lives.

By unofficial reckoning, the war has cost the Russian treasury upward of $5 billion to date in direct operating expenses, not counting the additional cost of aircraft and vehicles lost or damaged.[43] VVS activities in conjunction with the war diverted fuel paid for by appropriations originally intended to support VVS-wide continuation training. The war cut deeply into the VVS's fuel and munitions reserves, with no hope of near-term replacement. The head of the defense ministry's budget department acknowledged that the war's combined costs were not anticipated and that as one result, every second servicemen was forced to forgo a paycheck in July 1995. The net effect was to exacerbate an already dire funding crisis. Ministry of Defense arrears in wages and debts to suppliers now total over $2 billion. This has forced the ministry to take the extraordinary step,

---

[41]One such allegation was that a Su-27 was downed by a U.S.-made Stinger on November 29, even before the Russian ground invasion had begun. This is doubly implausible, since the Su-27 is a fighter-interceptor, not a ground-attack aircraft. Its restriction to a medium-altitude CAP above 20,000 ft would have kept it well outside the lethal envelope of the Stinger, even if the latter had been available to the Chechen rebels. See ITAR-TASS report, "The Pentagon Checks Whether There Were Stingers in Chechnya," December 14, 1994.

[42]Interfax, Moscow, August 19, 1995.

[43]See "Dead or Alive," *The Economist* (London), April 27, 1996, p. 54.

"impossible in any other state," of seeking bridge loans from commercial banks.[44]

The chief of the General Staff, Army General Mikhail Kolesnikov, candidly admitted that whereas the cost of such an operation in days past would scarcely have been felt by the High Command, "the conflict in Chechnya has required the concentration of our entire financial potential there." He added that "so far, we have not managed to get full reimbursement from the federal budget for the expenses we have incurred."[45] The practical outlook for such reimbursement is nil.

Grachev confirmed that in going into Chechnya, the Russian military was forced to rob Peter to pay Paul: "For the most part, funds appropriated for defense are being spent in Chechnya for munitions, food and clothing for servicemen, and fuel and lubricants. Many assets are being siphoned off there, creating a negative impact on the status of units not fighting in Chechnya."[46]

The war further aggravated an already low state of military morale. One Western reporter noted the deep bitterness of Russian soldiers in Chechnya and their "almost universal willingness to express that anger to foreigners with notebooks in their hands." Said a 20-year old tank gunner who had asked the interviewer to take a souvenir snapshot of him: "This way I can prove that I was part of the stupidest campaign of our time. If I live, I can show this picture to my

---

[44]ITAR-TASS, August 2, 1995. Cost estimates for the war have varied widely. Andrei Illarionov, director of the Economic Analysis Institute, wrote that Russia was spending the equivalent of $60 million a day at the height of combat operations and had exhausted $5 billion by the end of February. ITAR-TASS, Moscow, March 2, 1995. Other sources saw Russia's state coffers being drained by $30 million a day and the operation costing between $2 and $5 billion through early January. See Lee Hockstader, "Chechnya Draining Russian Economy," *Washington Post*, January 9, 1995. The total Russian allocation to defense for 1995 was only $11–$14 billion, barely a twentieth of the American defense budget.

[45]General Mikhail P. Kolesnikov, "Military Organizational Development as an Inalienable Part of the Establishment of Russian Statehood," *Krasnaia zvezda*, September 30, 1995.

[46]Interview by Andrei Vandenko, "Pavel Grachev: Yet More Dots Over Yet More I's," *Moskovskaia pravda*, August 2, 1995.

grandchildren and tell them how the Russian army was destroyed."[47] Not surprisingly, corruption and cynicism have spread rapidly as a result of the breakdown of morale and discipline. Chechen fighters have boasted on numerous occasions that they have been able to buy weapons from disgruntled Russian officers and conscripts.

Morale and motivation among Russia's aircrews were reportedly better. General Deinekin said that "it is not easy for the pilots there," but that "they are fulfilling their duties. We have not had a single desertion among the soldiers or officers in the force."[48] One account, however, indicated that some unit commanders flatly refused to allow their regiments to participate in the bombing of Grozny.[49] On reflection, such recalcitrance is understandable. Unlike the army, which had been called on repeatedly to deal with domestic political and civil unrest, this was the first time Russian airmen had been directed to commit violence against a designated foe on Russian soil. By all indications, most followed their orders with dispatch. It would not be surprising, however, if many did so with great reluctance. Only time will tell how the professionalism and self-respect of Russia's aviators will be affected by the Chechnya experience, particularly by the high incidence of both noncombatant fatalities and losses to friendly fire stemming from VVS actions.

Finally, the war further embarrassed an officer corps already humiliated and deeply riven at all levels. In the apt words of one American analyst, "nearly every aspect of military activity—from training, supply, coordination among services, strategy, tactics [to] morale and fighting spirit—failed the test of battle, feeding a growing resentment among the military leaders toward defense minister Pavel Grachev and, more serious still, increasing the potential for a breakup of the armed forces into feuding factions."[50] Among the

---

[47]Michael Specter, "Killed in Chechnya: An Army's Pride," *New York Times*, May 21, 1995.

[48]Vremya television report, June 28, 1995. This despite the persistently abysmal quality of life in the VVS, where more than 5000 rated officers have no apartment accommodations for themselves or their families.

[49]Mark Galeotti, "Decline and Fall: Moscow's Chechen War," *Jane's Intelligence Review*, Vol. 7, No. 2, February 1995, p. 51.

[50]Timothy L. Thomas, "Fault Lines and Factions in the Russian Army," *Orbis*, Fall 1995, p. 531.

problems and fault lines aggravated by the war, this account noted, were the emergence of rivalries between elite and regular components of the armed forces; the rise of parallel services such as the Federal Counterintelligence Service (FSK) and MVD troops in direct competition with the Ministry of Defense for funds and missions; a degeneration of regional military districts into de facto warlord enclaves; a dramatic decline in overall readiness; and the politicization of the military, as reflected in the participation of serving officers in organized factions—and more than 120 actively campaigning for seats in the State Duma during the run-up to the 1995 parliamentary election.

KEY ACCOMPLISHMENTS

The war in Chechnya was a tailor-made test of Russia's new regional security doctrine. By most indicators, the armed forces fell far short of a respectable showing in it. Nevertheless, the experience—even as a negative one—bore out the essential correctness of the Russian military's power-projection emphasis, for Chechnya proved a convincing prototype of the sort of security challenges Russia is likely to confront around the former Soviet periphery at least in the near term. It further reaffirmed what the VVS has already come to recognize, namely, that its greatest acquisition need for the coming decade is not new combat aircraft but more airlift.

On balance, the VVS acquitted itself better in Chechnya than Russia's ground forces by a considerable margin. Informed U.S. government analysts give it "a passing grade for a credible performance, even though saddled with a poor military plan."[51] In particular, the VVS ran a professional airlift operation that, by all signs, met the needs of ground commanders handily. It also, albeit with some egregious lapses, did better than either the ground forces or the higher military leadership in owning up to its deficiencies and failings. Such willingness to be self-critical will be essential if the VVS is ever to rise above the corruption nurtured by the Soviet system and recover to a state of good institutional health.

---

[51] *Aviation Week*, August 7, 1995.

A Validation of Airlift

Although badly truncated in the wake of the USSR's disintegration, VTA performed effectively in Chechnya and deserves high marks for doing as well as it did under conditions of great adversity, including a need to pull together assets from widely dispersed locations throughout Russia. Despite frequent foul weather, a severe shortage of spares, and reduced aircrew proficiency owing to curtailed annual flying, it maintained a high sortie rate throughout the initial buildup without losing a single aircraft. In all, VTA reportedly moved over 20,000 troops and 1000 units of Russian hardware into the theater. With a reported total of 40,000 troops committed altogether at the height of Russia's involvement, that adds up to about half of all Russian ground forces deployed.[52] From the end of November 1994 through February 1995, VTA transports flew a total of 492 reported sorties for 4020 flying hours, during which 22,000 men, 1140 vehicles, and 3057 metric tons of cargo were delivered.[53]

As noted in Chapter Three, an analysis by the VVS's Central Research Institute (NII) had predicted that, with its reduced assets, VTA would have to commit virtually all its serviceable transport aircraft to move just a single airborne division in two sorties.[54] Indeed, VTA's funding had been cut back so severely that it had been driven into the commercial airlift business in pursuit of nonbudgetary income to sustain its operations. Yet despite this acknowledged handicap, the NII's assessment proved overly pessimistic. VTA ran an efficient operation into Mozdok and Vladikavkaz by handling its problem piecemeal, using a mixed fleet of Il-76 Candid, An-12 Cub, and An-22 Cock transports. Early on, the An-124 Condor was used as well to transport 1000 commandos and their armored personnel carriers from the 104th Guards Airborne Division in the course of a 24-hour period.

---

[52]Ruslan Ignatiev, "Nonflying Weather Sets In," *Rossiiskaia gazeta*, March 4, 1995.

[53]*Aviation Week*, August 7, 1995.

[54]See "Russian Air Strategy and Combat Aircraft Production: A Russian Air Force View," in Randall Forsberg, ed., *The Arms Production Dilemma: Contraction and Restraint in the World Combat Aircraft Industry*, Cambridge, Massachusetts, MIT Press, 1994, p. 45.

The Beginnings of Candor in Self-Assessment

The war in Chechnya also revealed the Russian military, or at least important parts of it, to be refreshingly honest with itself in frankly admitting to its shortcomings as well as congratulating itself on its successes. This was in sharp contrast with the High Command's unseemly reaction following the lopsided fighting between Israel and Syria over Lebanon's Beka'a valley in 1982, when the Soviets proved incapable of comprehending the main implications of the failings displayed by their Syrian clients.[55] It also contrasted with the Soviet military's reluctance to face up to its own combat failings in Afghanistan, and for much the same reason: To have done otherwise would have required the Soviets to concede their fallibility.

The Russian military has a way to go yet, however, before it can be said to have completely unburdened itself of its former Soviet habits. Once the chrome had been knocked off the halo of Soviet communism by 1991, Russian defense professionals were freed to do an objective job of assessing combat operations on both sides in the Persian Gulf War, which they did with remarkable insight and intellectual acuity.[56] These same professionals have shown a less uniformly impressive record, however, at making useful sense of their own subsequent combat experience. Because of the humiliation caused by Russia's debacle in Chechnya, at least some are having, once again, a hard time facing up to combat facts.

Typical of the post hoc rationalization and excuse-making apparent in some quarters was the contorted reaction of the deputy head of the General Staff's Military Science Directorate regarding why things went so wrong in Chechnya. Starting off on the wrong foot, he said: "Under no circumstances should events in the Chechen Republic be considered combat operations in the classic sense."[57] In near-

---

[55]See Benjamin S. Lambeth, "Moscow's Lessons from the 1982 Lebanon Air War," in Air Vice Marshal R. A. Mason, ed., *War in the Third Dimension: Essays in Contemporary Air Power*, London, Brassey's, 1986, pp. 127–148.

[56]For detailed amplification, see Benjamin S. Lambeth, *Desert Storm and Its Meaning: The View from Moscow*, Santa Monica, California, RAND, R-4161-AF, 1992.

[57]Interview with Major General Yevgeny Nikitenko, "Grozny Is Not the Desert: First Attempts to Derive Lessons from the Army's Actions in Chechnya," *Krasnaia zvezda*, January 27, 1995.

flawless Orwellian doublespeak, he added that Russia's forces "are not conducting a military operation as such," explaining that the sort of incursion conducted in Chechnya "does not figure in basic military texts." Accordingly, he wrote, "no one ever formulated for the military the procedure, methods, and means of conducting such an operation."

Admitting that a "miscalculation" had been made in anticipating the scale and intensity of rebel resistance, General Nikitenko suggested that the hurried decision to send tanks into Grozny without infantry cover was based on "a glimmer of hope" that the rebels had only a limited number of ATGMs, which "regrettably did not prove to be the case." He implied that had the High Command had its way, a fundamentally different force would have been sent in, organized along the lines of allied formations in Desert Storm. Chechen civilians would have been "asked" to leave the zone of impending operations, and Russian ground forces would have entered Grozny only after "massive bombardments" by Su-24Ms and Su-25s to "completely destroy the mini-army that Dudayev had set up." Using "all the might available to the Russian armed forces," he added, "there would not be even a shadow of Dudayev's cutthroats left on the face of the earth. At the same time, they could not have done any palpable damage to our troops." This is not learning; it is non-recognition and denial.

Russia's military leadership has gone so far as to forgo including any putative teachings of the Chechen war in the studies curriculum at the General Staff Academy and at other senior service schools, on the dubious ground that the conflict was "atypical" in having been conducted on Russian soil.[58] Grachev chaired a narrowly focused "lessons learned" session among technical specialists in February 1995 aimed expressly at a postmortem on the combat employment of armor. He concluded that Russian tanks showed "excellent battle-

[58]Colonel (Ret.) Vitaly V. Shlykov, "The War in Chechnya: Implications for Military Reform and the Creation of Mobile Forces," paper presented at the Third Annual Conference on Russian Defense Decisionmaking," U.S. Naval Postgraduate School, Monterey, California, November 7–8, 1995, p. 12.

field characteristics," notwithstanding their misuse in being committed in ones and twos without protective cover.[59]

With greater insight, the Russian airborne commander, Colonel General Yevgeny Podkolzin, remarked that the first mistake was the government's failure to prepare public opinion, closely followed by the decision to begin the operation at a time of year when second-year servicemen were being discharged and new conscripts inducted. Yet a third needless complication was the prohibitive weather known to afflict the North Caucasus region in November through January: "In these conditions," he conceded, "it is impossible to send your aviation up. Or if it does get airborne, it has to stay at high altitude, which naturally makes it hard to fulfill its missions." He added that the American leadership wisely spent a solid six months getting its citizens ready for Desert Storm.[60]

General Kolesnikov admitted that the Chechnya experience "graphically illustrated a large number of problems that have built up in the military." He said the war provided "food for thought and a basis for making certain changes in operational tactics, especially with respect to urban combat, the organization of communications, psychological training, and tactical interplay between units." He added that most of the difficulties spotlighted were simply a consequence of the dire funding situation and that the problem "cannot be resolved without a substantial economic upsurge in Russia."[61]

---

[59]See "Russian Military Assesses Errors of Chechnya Campaign," *International Defense Review*, No. 4, 1995, pp. 5-6.

[60]Interview with Colonel General Ye. Podkolzin, "Don't Cry Over the Military! We Have a Military. And Russia Has Defenses! We Were Marginalized," *Sovetskaia Rossiia*, February 23, 1995.

[61]Interview with Colonel General Mikhail Kolesnikov, "Despite All the Difficulties, the Russian Military Has Stood Its Ground and Remains a Guarantor of Stability," *Krasnaia zvezda*, May 6, 1995. General Kolesnikov has worked hard to distance himself from the Chechnya debacle. He has kept a low profile, refrained from endorsing the intervention, and generally avoided commenting on it in the press. This silence suggests a serious rift between Kolesnikov and Grachev and that the former is protecting his bona fides with the High Command. On May 7, 1995, Kolesnikov was promoted to the rank of General of the Army, making him one of only two Russian four-stars at the time and the equivalent of Grachev. A little more than a year later, after placing first in the initial round of the 1996 presidential election, Yeltsin fired Grachev and replaced him several weeks thereafter by Colonel General Igor Rodionov, the commandant of the General Staff Academy. In the interim, Kolesnikov was appointed acting defense minister.

With conspicuous exceptions dictated by the Yeltsin government's determination to cover up its worst excesses in Chechnya, the VVS was generally candid about the problems its air campaign revealed. By the admission of its own airmen, the performance of the VVS in Chechnya "exposed shortcomings in the combat training of Frontal Aviation pilots," many of whom had to "regain lost skills in the midst of military operations." This deficiency was described by one serving officer as partly the natural result of a long-standing fixation on flying safety at any cost, which dated back to the introduction of the second-generation MiG-21 into service during the mid-1960s: "Where this led to can be seen from the present state of tactical air training. Pilots complain of numerous restrictions in practicing difficult aspects of training sorties." According to this account, the unstated but still-binding watchwords of the VVS's leaders remain: "Take no risks, do not complicate, and avoid innovation."[62]

General Deinekin, a bona fide air power professional, has stressed more than once that "honesty in aviation is an absolute must."[63] Partly as a result of this outlook, his commentary on air operations in Chechnya has been factual and generally frank, in marked contrast with previous Soviet practice. He was forced onto the defensive early by a barrage of media allegations that the VVS was indiscriminately bombing Chechen hospitals, schools, and residential areas. One press account characterized his aircraft as "almost haphazardly bombing a nearly defenseless city."[64] A Moscow television station reported a still-burning flare that had come down by parachute as a Russian bomb that had gone off in a Grozny housing project.

In the worst cases, the VVS was accused, sometimes justifiably, of outright falsification about its activities. For example, the VVS initially denied that it was involved in air operations in Chechnya at all. Only on December 5 did Grachev admit that Russian combat aircraft had bombed the airfield in Grozny. On December 23, the respected human rights envoy Sergei Kovalev charged that government repre-

---

[62]Lieutenant Colonel Valery Veshnikov, "If Flying, Then How to Fly, If Shooting, Then How to Shoot. . .," *Armeiskii sbornik*, No. 7, July 1995, pp. 26–27.

[63]Ibid.

[64]Steven Erlanger, "Russia's Army Seen as Failing Chechnya Test," *New York Times*, December 25, 1994.

sentatives the previous night had "claimed that the night bombing raids on Grozny had stopped and that the explosions in the town were initiated by Dudayev himself. That is a blatant lie. . . ."[65] There was also a question a week and a half later as to whether the VVS's bombing of Grozny had been halted in accordance with President Yeltsin's January 4 edict to that effect. Pointedly upbraiding Grachev on this score, Yeltsin openly admonished him at a Security Council meeting: "And I want to hear absolutely precise information from the defense minister."[66]

The VVS was also evasive about the participation of Long-Range Aviation (LRA) in the war. General Deinekin insisted that only Frontal Aviation ground-attack aircraft and army helicopters took part in actual combat operations. Senior VVS officials denied that LRA strategic bombers were used to bomb cities.[67] Official statements claimed that the Tu-22M3 Backfire was used only for night flare drops and for dropping propaganda leaflets over Grozny.[68] Foreign reporters alleged, however, that the Backfire was also used on several occasions to bomb Chechnya's forces directly.

General Deinekin appeared genuinely distressed at allegations that his pilots had intentionally bombed noncombatant civilians, calling such accusations "an evil fabrication." In the face of such charges, he launched a post-attack reconnaissance mission on December 29, the results of which reportedly "confirmed" that VVS strikes had been directed solely against military targets. These were said to have included a Chechen tank repair facility, troop marshalling areas, and the presidential palace.

Deinekin later complained that because of such alleged slanders, "the public is set against the Russian armed forces and their aviation," while "absolutely no attention is paid to the pilots' arguments,

---

[65]Statement to the radio station *Ekho Moskvy*, December 23, 1994.

[66]Carey Goldberg and Sonni Efron, "Yeltsin Demands Withdrawal Date from Chechnya," *Los Angeles Times*, January 7, 1995.

[67]Said one flatly: "There are no area targets to be destroyed by heavy aviation in the theater of operations in Chechnya." Interfax, Moscow, June 2, 1995.

[68]Interfax, Moscow, March 3, 1995. See also the interview with General Deinekin, "Official Version: 'We Only Attacked Grozny Military Installations . . .,'" *Trud*, January 17, 1995.

and they are blamed for virtually all the misfortunes of the war. . . . They had to operate against what was virtually a full-scale army, armed to the teeth. . . . Quite often tanks and guns were set up not in an empty field but near schools and kindergartens and in the yards of apartment compounds. . . . If the air force had not fulfilled its missions, the number of dead Russian soldiers would have been far greater. We have every ground for taking pride in our fliers' courage."[69]

Because of complex front lines and the inaccuracy of radar bombing through clouds (with average miss distance no better than 450 ft), Russian officials do not deny that some VVS bombs landed on Russian troops. General Deinekin was uncomfortable with charges to this effect, and he repeatedly defended the professionalism and discipline of his pilots. However, on one occasion in early January, he conceded that he could not rule out inadvertent fatalities, as well as the destruction of civilian apartment compounds through accidental stray bombs.[70]

Deinekin also acknowledged the "many conflicting assessments" of his air force's performance in Chechnya. Yet he insisted that notwithstanding objective difficulties, Russia's pilots "fully coped with their missions, demonstrating the high effectiveness and reliability of Russian weapons and aviation equipment and their own high skills."[71] Giving credit where it was due, he also conceded that the Chechen rebels were as effective as they were because they had received the same training and used the same equipment as Russian forces.

General Deinekin was so concerned to correct the bad press the VVS received after its initial poor showing in Chechnya that he staged a firepower demonstration at a weapons range not far from the war zone for the air attachés from 40 foreign nations. He was unambigu-

---

[69]Interview with Colonel General Petr S. Deinekin, "The Air Force in the Chechen Conflict," *Krasnaia zvezda*, March 17, 1995.

[70]Sergei Ovsiyenko, "Low Cloud Cover Screens the Air Force: Commander in Chief Petr Deinekin Is Unable to Dispel the Haze Over Air Force Actions in the Chechen Conflict," *Rossiiskiye vesti*, January 13, 1995.

[71]Interview with Colonel General Petr S. Deinekin, "Flying in Your Dreams and in Reality," *Rossiiskiye vesti*, August 17, 1995.

ous about his motives in this respect: "A powerful propaganda ploy was recently organized in the press in connection with events in Chechnya. Its intent was to prove that our aviation is not capable of using precision weapons, that our bombs hit hospitals, markets, and children's homes. Of course, it was hard for us to endure this, so we decided to show the military intelligence people and attaches representing other countries in Russia what our Russian aviation is capable of."[72]

To provide a suitably impressive venue, Deinekin picked the VVS high test range in Kabardino-Balkaria, situated at an elevation of 9000 ft mean sea level in the Elbrus mountains. The demonstration, most likely performed by the VVS's most proficient aircrews from Lipetsk, included a simultaneous launch of two precision-guided air-to-ground missiles.[73] Afterwards, Deinekin said: "I'm very satisfied. The pilots have shown flying skills of the highest class. Despite the marginal weather, all landed safely."[74] He later reiterated that the demonstration had been laid on to counter adverse comment about Russia's armaments intended to "shatter their glory," adding that "today we showed the best our aviation industry has."[75]

PROBLEMS AND LESSONS INDICATED

There are few profound learning points to be drawn by the VVS from its experience in Chechnya, since so many of the problems dramatized by its uneven performance reflect nothing more complex than the severe funding shortage that has afflicted it since the USSR's collapse. If anything, the main lessons amounted simply to worst-fears-confirmed about the VVS's eroded capabilities. What the war did, in the final tally, was to bathe the VVS's problems in the cold light of reality and to identify beyond question the hurdles that remain ahead on the road to recovery.

---

[72]Interview with Deinekin, "Flying in Your Dreams and in Reality."

[73]Pavel Anokhin, "A Military Landing in the Elbrus Region," *Rosiiskiye vesti*, June 6, 1995.

[74]Vremya television report, June 28, 1995.

[75]Interfax, Moscow, June 1, 1995.

The following discussion will not try to account for VVS "lessons learned," for two reasons. First, such lessons are rarely self-evident to the outside observer. More important, such an approach all too often assumes—wrongly—that the adduced "lessons" have in fact been understood and assimilated. A more useful approach is to think less definitively in terms of "lessons indicated." Four such sources of reflection, in particular, have no doubt captured the attention of the VVS's leaders in the wake of their recent combat trial in Chechnya.

The Burdens of Bad Planning

The VVS found itself up against needlessly high odds from the outset in Chechnya. To begin with, the weather deteriorated at precisely the time the initial ground invasion began. Defense minister Grachev, moreover, underestimated the fighting capacity of the resistance. Finally, Russian ground forces failed to encircle the capital city of Grozny before entering, thereby allowing in enemy reinforcements and later enabling many irregulars to escape and continue fighting from the surrounding hills. There is nothing that air power could have done to compensate for these flatfooted miscalculations.

By all accounts, the decision to commit Russian troops to the invasion was made within Yeltsin's inner circle, without consulting of the senior military leadership. Grachev willingly, even reflexively, acceded to this decision, yet failed to carry out any prior assessment of the situation or to prepare his forces for what was to come. There was no apparent concept of operations behind the incursion beyond a vaguely defined injunction to "disarm illegal formations" and to lend fire support to MVD troops. Grachev had assured his superiors that his army would "cleanse the city of rebels" by week's end.[76]

---

[76]Quoted in Carey Goldberg and Sonni Efron, "Russians Suffer 'Crushing' Loss in Rout by Chechens," *Los Angeles Times*, January 3, 1995. The well-informed Russian defense reporter Pavel Felgengauer highlighted the fundamental error of letting tank convoys into the city without first sanitizing the area. He noted openly: "It is very strange that the military leadership, primarily General Grachev, as a representative of the military's professionals, could not say 'no' to the politicians." "Russia on the Brink of a Catastrophe: The Russian Subunits Which Entered Grozny Have Been Routed," *Segodnya*, January 5, 1995.

Indeed, on November 29—before the botched invasion two weeks later—he boasted that a single airborne regiment could take Grozny in two hours.[77] The plan was for a quick Russian advance into Grozny, after which resistance from Dudayev and his forces was expected to collapse.[78]

The VVS can hardly be blamed for the uncooperative weather during the opening phase of the invasion, which all but precluded effective air-to-ground operations. Even the following February, General Pavlov stated that 95 percent of the month was nonflyable by normal peacetime training rules because of meteorological conditions below First-Class pilot minimums. Any air force, including the USAF, would have been similarly constrained in such circumstances. Indeed, winter weather hampered allied air operations on repeated occasion during the Persian Gulf War, often nullifying the capabilities of the coalition's precision-guided weapons. It has had a similar adverse effect on the more recent NATO air activities over Bosnia.[79]

The invasion of Chechnya was sharply scored by some of Grachev's most senior subordinates, including his deputy minister, General Boris Gromov, and Lieutenant General Alexander Lebed.[80] Gromov complained that the operation was planned in "profound secrecy"

---

[77]See Wendy Sloane, "A Goliath No More: Russian Army Takes It On the Chin," *Christian Science Monitor*, January 13, 1995.

[78]One Western account reported that the VVS had targeted Dudayev's family compound in the suburb of Tashkala and Grozny's electric plant, but missed: "Instead, they scored direct hits on such objects as the muddy back yard of pensioner Yevgeniya Pogosian." Carey Goldberg, "Russia Steps Up Bombing Near Chechen Capital," *Los Angeles Times*, December 20, 1994.

[79]A major difference is that poor weather in the target area during Operation Desert Storm typically forced a mission abort because of strict rules of engagement prohibiting weapon release unless the prebriefed target could be acquired and positively identified. In Chechnya, the VVS evidently dropped without regard for the possibility of collateral damage.

[80]The operation also prompted early recalcitrance from senior commanders on the scene, one of whom, Major General Ivan Babichev, halted his advance and refused to fire on unarmed civilians. See Alessandra Stanley, "Russian General Halts His Tanks in Chechnya in Sign of Uneasiness," *New York Times*, December 17, 1994. General Gromov stated frankly in an interview that that the operation was "being handled by idiots." Interview by Livia Kling, "Idiots Are Responsible for This Operation," *Kurier*, Vienna, January 5, 1995. Lieutenant General Leonid Ivashov, echoing this judgment, said that "another major act of stupidity has been perpetrated."

and that no one on the military collegium was consulted on it.[81] Another critic flatly charged that in acceding to such a hastily contrived operation, Grachev was following "principles of political expediency rather than military science."[82] By this same account, the defense ministry's intelligence reporting on the state of Dudayev's forces was badly in error. Russia's troops were accordingly unprepared, their ingress routes had not been properly secured, and timely measures to neutralize Dudayev's forces had not been undertaken. One anonymous army officer pointed out that many of his superiors regarded Grachev as a "weak, incompetent minister with the mentality of a commander of a troop division rather than of a minister." This critic faulted Grachev for "surrounding himself with an entourage of dull but loyal hacks" and disparaged him as one who only "manages to hang on because of his loyalty to his patron."[83]

Those in Yeltsin's kitchen cabinet who elected to initiate this war could not have picked a worse time of year from a weather perspective. But an even bigger mistake was to drive into the center of Grozny with tanks and armored personnel carriers exposed to sappers hidden inside and atop buildings, having failed first to encircle the city, clear an ingress route, and secure a safe escape option. Grachev sent in some 250 unprotected tanks and armored personnel carriers. These were quickly bottled up and decimated by Dudayev's irregulars. The Russian tank crews had received little prior training worthy of the name, and no exposure whatsoever to the most elementary principles of urban warfare. Nor did they show any signifi-

---

[81]Interview by Aleksandr Zhilin, "Boris Gromov: The Operation Was Prepared in Profound Secrecy . . .," *Moskovskiye novosti*, No. 1, January 8–15, 1995.

[82]Sergei Surozhtsev, "The Legendary Army in Grozny: The Opinion of a Military Expert on the Actions of the Russian Army in Chechya," *Novoye vremya*, Nos. 2–3, January 1995, pp. 14–15.

[83]"Letter from Officer X," *Time*, January 23, 1995. One of the Russian generals who quit rather than lead troops into Chechnya later opined that Grachev lacked the courage to tell Yeltsin that his troops were unprepared for the invasion. See Fred Hiatt, "Russian General Assails Defense Minister on Chechnya," *Washington Post*, January 27, 1995. Another account noted that in his previous incarnation as airborne commander, Grachev had been known to try to impress important visitors by ordering his paratroopers to jump in excessively high wind conditions, resulting in injuries on landing. This commentator noted that Grachev's seeming readiness to endorse a half-baked invasion plan merely to please Yeltsin would have been "in character." Reported in Sonni Efron, "Army In Tatters May Threaten Russian Reform," *Los Angeles Times*, February 27, 1995.

cant degree of coordination, since they had been cobbled together only days earlier from often widely dispersed units. Russia's poor performance further reflected the fact that most conscripts had less than a year's service time, according to the ground forces commander, Colonel General Vladimir Semenov.[84]

Because of poor planning, the invading tanks became separated from supporting infantry almost immediately. This made them easy prey for Chechen snipers armed with ATGMs and rocket-propelled grenades (RPGs). The invasion repeated almost to the letter the same errors that had been committed during the earlier November 26–27 incursion, only at a significantly higher cost in Russian casualties. Only after this debacle did Grachev call in elite detachments from Yekaterinburg and from the Tula airborne division.

Regarding the army's sorry showing in Grozny, Lieutenant General Aleksandr Lebed remarked that amazing though it sounded, "all the mistakes Soviet troops made in Afghanistan have been repeated in Chechnya. The army totally ignored local conditions, religion, and customs. No one planned the operation. It was started 'Russian style' on the off-chance that it would work."[85] For its part, the VVS played the hand it was dealt by defense minister Grachev. Even had the ground campaign been conducted with greater forethought, effective air preparation would have been difficult to impossible because of the built-up urban setting of the fighting.

The Price of Financial Starvation

The VVS also felt the effects of the curtailed funding for operations and training it had been forced to endure since the collapse of the Soviet Union. In Operation Desert Storm, the United States and its coalition allies saw the payoff of fifteen years of prior intensive training and experience at mission planning, force integration, and combat employment. In Chechnya, the VVS saw the results of the absence of these crucial equities. Both examples, in their way, strongly reaffirmed the long-standing axiom among airmen that, for

---

[84]Interfax, Moscow, February 23, 1995.

[85]Quoted in Michael Specter, "For Russia's Army, Humbling Days," *New York Times*, January 8, 1995.

better or for worse, you fight like you train. The war in Chechnya came close on the heels of a plaintive lament by Grachev to parliament that "not a single army in the world is in such a catastrophic state" as Russia's. Grachev warned the legislators that without supplemental funding "the irreversible process of losing our capability will occur, and the armed forces will simply collapse."[86]

Low proficiency as a result of reduced flying hours and curtailed training opportunities was a serious problem for the VVS. General Deinekin frequently conceded the impaired readiness of his aircrews owing to training cutbacks caused by lack of fuel. He stated that some of his "pilots" (intentionally setting the word off in quotation marks for effect) were averaging only 15 hours a year, in bleak contrast to the Western norm of 180–240 hours annually. He also disclosed that VVS pilots with night currency were "few and far between" and admitted that the toughest challenge was often simply to find pilots who would not collide with each other in midair. Because of curtailed training, the VVS is steadily losing its cadre of First-Class pilots, who must meet annual currency minimums in both flying hours and mission events to retain their top aeronautical ratings.

It turned out that weapons instructors from the VVS's combat training and aircrew conversion center at Lipetsk and test pilots from the military flight test center were the only pilots proficient enough to use precision-guided munitions (PGMs) in combat. They were accordingly pressed into service for most PGM attacks in Chechnya. This was not openly dwelt upon by the VVS, for understandable reasons. But it was alluded to on at least one occasion.[87]

---

[86]Quoted in Steven Erlanger, "Dire Warning to Legislators On Plight of Russian Army," *New York Times*, November 19, 1994. This was contradicted by assurances the same month by Grachev to Yeltsin that despite all, the armed forces were "fully combat ready and capable of carrying out any task." This rosy picture was sharply contested by a formal reclama to the parliament by eleven generals from the Ground Forces' Military Council, headed by Colonel General Semenov, complaining that there had not been a division-level training exercise since 1992 and that a third of the army's helicopters were nonflyable. See Igor Chernyak, "Scandals: Infantry Generals Attack Grachev," *Komsomolskaia pravda*, December 10, 1994.

[87]One report indicated that VVS test pilots from Akhtubinsk who delivered PGMs in Chechnya had been awarded Philips television sets. By comparison, said the report, "General Klishin's ace pilots were usually given a wrist watch—or, at best, a domestically made television set—for testing state-of-the-art aircraft." Lieutenant General Yury Klishin commands the State Flight Test Center at Akhtubinsk. See Anton

Ground forces aviation experienced similar consequences from the preceding four years of curtailed funding for operations and support. Its commander, Colonel General Pavlov, stated that 59 percent of his helicopter aircrews were Afghan veterans and that all his aircraft commanders flying in Chechnya were First-Class rated pilots. Few, however, were current in night/adverse weather operations or weapons delivery. First-Class pilots were cleared to fly to minimums of a 300-ft ceiling and a half-mile visibility. Most, however, had flown only 40–50 hours during the preceding year, barely a third their peacetime training norm. They also complained of operating old equipment. The average service time on the Mi-24 airframes employed in Chechnya was more than 15 years. Most had previously seen hard use in Afghanistan. To add insult to injury, virtually everything needed to support rotary-wing operations in the war was in short supply. General Pavlov said his staff had to scour the entire country to scrounge enough flak jackets and flight helmets for his helicopter crews.

General Pavlov also said the limited annual flying hours allotted to his pilots had approached a "danger threshold," adding that "with this amount we will only be able to maintain a set level of combat readiness for one or two years. But after the departure of those fliers with 10 to 15 years of flight experience on their shoulders, these '48-hour kids' will never be fully ready for combat."[88]

Disjunctions in Joint Force Integration

Attack helicopters were often used in conjunction with ground attack aircraft. There were also combined-arms operations pitting Russian artillery, multiple-launch rockets, attack helicopters, and fixed-wing jets against rebel targets. Coordinating attack helicopter operations with infantry and armor proved more than once to be a problem. Lessons learned the hard way in Afghanistan were frequently forgotten, making it necessary, by one account, for Russian forces "to step on the same rake again."[89]

Vasiliyev, "Pilots Given a Television Each for Grozny Bombardments," *Novaia yezhednevnaia gazeta*, January 14, 1995.

[88]Beaver, *Jane's Defense Weekly*, January 10, 1995, p. 79.

[89]Surtsukov and Prokopenko, *Krasnaia zvezda*, July 18, 1995.

General Lebed, who did not participate in the fighting but was well situated to know what was happening, remarked afterward that in Chechnya "planes fly on their own, guns fire on their own, without coordination, and no one can come to any agreement with anyone even theoretically, because each has his own boss."[90] This suggests that there was no on-scene commander in chief with clearly subordinated joint force component commands like the arrangement from which the allied coalition took its tasking in Desert Storm. More likely, the component commanders, in classic Soviet fashion, reported separately to, and took directions from, the High Command in Moscow—in this case General Grachev.[91]

Russian attack helicopter operations, in particular, showed the effects of a less-than-seamless integration with the ground forces. Rotary-wing aviation, previously assigned to the VVS, was transferred back to Soviet army ownership in 1990. The VVS commander in chief at that time, Colonel General Shaposhnikov, noted that there were no real alternatives to this, considering that attack helicopters, as a prime antitank asset, would immediately be remanded to combined-arms commanders in case of war. This prompted concerns among members of the helicopter community that they might find themselves forgotten stepchildren of the ground forces, commanded by dilettantes untutored in flight operations, and stuck last in line for such amenities as food, housing, and uniforms.

As it turned out, all the Russian combat arms suffered the deprivations of post-Soviet defense budget cuts. But the helicopter community proved prescient with respect to the first concern. As for the possibility of incompetent directives from unschooled ground-force commanders, Shaposhnikov insisted that, although ground commanders would assign the mission, "the process of carrying it out will

---

[90]Interview on the Moscow Mayak radio network, October 16, 1995.

[91]By one account, Colonel General Deinekin was designated Senior Controller for Air Operations, with the VPVO's commander in chief, Colonel General Prudnikov, subordinated to him as Controller for Air Defense Support (see Richard Woff, "Who's Who in the Chechen Operation," *Jane's Intelligence Review*, Vol. 7, No. 4, April 1995, p. 161). Whatever the actual relationship between the two service chiefs may have been, there is no evidence that either was ever actually on site at the operational command center in Mozdok.

be developed by the aviation staff."[92] Things did not always work out so smoothly in Chechnya, however.

A particular difficulty, in the words of one helicopter pilot, was that many combined-arms commanders had only a "vague idea" of the combat capabilities of helicopters and of restrictions on their use with respect to weather, aircraft and weapons limitations, weapons range, airspeeds, and aircraft load-carrying capacity. "What they are *not* short of," he added wryly, "is resolve. And if aircraft are subordinated to such a commander, it is tough to predict the consequences." Typically, an infantry unit commander would assign his helicopter pilots the mission of destroying a rebel tank in the streets of Grozny: "His logic was simple: 'You have ATGMs, and that means you can destroy it.' But suppose the tank's location was unknown? 'Fly along the street, then, and as soon as it appears. . . .' Actual combat, however, is not an American film hit. Sending a helicopter in on such an assignment means losing both the crew and the helicopter. Misunderstandings often arose because of a lack of coordinated operations."

The Limits of Air Power in Irregular War

The VVS had a golden opportunity in Chechnya to learn that air power cannot invariably work its reputed magic in circumstances where the target set is elusive, problems predominate in target location and identification, and there is an ever-present danger of unintended harm to noncombatants. In the comparable case of Bosnia, retired USAF General Charles Boyd, the former deputy commander in chief of the U.S. European Command, has written from first-hand involvement that "despite its appeal to the amateur strategist, a reliance on air power alone—the strike option—in this type of terrain with these kinds of targets has never held any real promise of conflict resolution." Although a robust use of air power, he suggested, might alter enemy behavior if applied without concern for civilian casualties, it can have no more than a near-term effect. It will also never vitiate the larger reality that allied strategy "cannot produce an enduring solution with military force—air or ground—only one that will

---

[92]Interview with Colonel General Ye. Shaposhnikov, "The Air Force Today and Tomorrow," *Aviatsiia i kosmonavtika*, No. 8, 1990, pp. 2–3.

last until it departs."[93] Among all possible lessons from the war, it is perhaps this one that Russian defense planners could most usefully ponder as they reflect on their grim experience in Chechnya.

IMPLICATIONS

Russia's war against Chechnya was emblematic of the security challenges the VVS is most likely to face in the decade ahead. The war was regional yet remote from the center of Russia. It featured a technologically unsophisticated yet almost fanatically determined ethnic opponent. It presented no air-to-air threat and offered a permissive operating environment for attack aircraft, except at low altitude, where widely dispersed antiaircraft guns and shoulder-fired infrared SAMs posed a constant danger. Finally, it entailed little air opposition and few target arrays, and accordingly did not place great demands on the VVS for high-technology performance. All in all, despite the occasional effective use of precision-guided weapons against key targets, quantity prevailed over quality in VVS operations in Chechnya.

A year and a half before, the VVS had conducted a rehearsal of sorts for its new role in Russian strategy in a two-day deployment of Su-24s, Tu-95s, and Tu-160s in a long-range strike exercise called *Voskhod '93*, described above in Chapter Five. As successful as that exercise proved to be, however, it was less typical of the real-world demands the VVS is likely to face in its future tasking than the war in Chechnya. The latter placed the greatest premium on airlift and sustainability for prolonged air support to ground operations rather than on rapid response, complex force packaging, and high-technology weapons employment. At least one unspoken motive behind the VVS's involvement in Chechnya may have been an urge to show that it could do anything the U.S. Air Force is capable of doing. General Deinekin spoke with measured respect for the "much-vaunted Americans," noting how almost inescapably "we always try to compare ourselves with them."[94] Since the end of the

---

[93]General Charles G. Boyd, USAF (Ret.), "Making Peace With the Guilty," *Foreign Affairs*, September–October 1995, pp. 37–38. This article was later translated and reprinted in the Russian military daily, *Krasnaia zvezda*.

[94]Interview by Vladislav Listyev, Ostankino First Channel, December 14, 1994.

cold war, Russians have repeatedly chafed at not being naturally accepted by the West as an equal power. They have also been sensitive about the allied coalition's bravura performance in the 1991 Gulf war, in considerable part against Soviet weapons and operating doctrine. In Chechnya, these sensitive Russians may have seen a passing opportunity to emulate at least America's largely painless intervention in Haiti.[95]

Any such hopes, however, fell short of being realized by the VVS's performance. General Deinekin characterized Chechnya as a "serious test of the Russian Air Force's combat capability."[96] He made a special point to portray VVS operations there as proof that the VVS is capable of a performance in waging air war comparable to the allied coalition's performance in Desert Storm. Yet the Chechnya war and the tasks that befell the VVS in conducting it mirrored neither the magnitude, the complexity, nor the character of the far more intensive and demanding Persian Gulf War. More important, it was not at all proven by Russia's experience in Chechnya that air power has yet emerged as the dominant force element in Russian defense planning. General Deinekin has insisted, probably correctly, that the VVS contains the intellectual cream of the Russian military. He has also pursued a determined effort to make the VVS the central force element in the armed forces.[97] Nevertheless, there was a predominant ground emphasis in the Chechnya campaign, and the war was planned and led by ground forces officers throughout.

That said, the VVS did remarkably well considering the many problems that have afflicted it since the USSR's demise. General Deinekin has freely admitted that the troubles he inherited when the Russian VVS was first established in May 1992 were mighty enough to "make his head spin."[98] In the ensuing four years these troubles have shown little sign of abating. By all indications, VVS operations in Chechnya were not constrained by a lack of fuel, munitions, or other

---

[95]See "The Chechen Trap," *The Economist* (London), January 7, 1995, p. 39.

[96]Colonel General Petr S. Deinekin, "Russia Has Been, Is, and Will Remain a Great Aviation Power," *Krasnaia zvezda*, August 19, 1995.

[97]Interview with Colonel General Petr S. Deinekin, "Russia Will Not Remain Without Wings," *Krasnaia zvezda*, November 5, 1994.

[98]Interview with Colonel General Petr S. Deinekin, "Pilots Are Always Equal to the Task," *Trud*, August 19, 1995.

consumables. General Deinekin conceded, however, that assets expended during the fighting "were replaced, to a considerable extent, not by deliveries from industry but by removal from the stocks of other air formations."[99] These material demands of the war in Chechnya set back an already strained VVS recovery effort by placing burdens on limited fuel and war reserve stocks that cannot be replenished.

More recently, the VVS was the instrument of choice that finally succeeded in taking out the Chechen leader, Dzhokar Dudayev, in a precision air attack on April 24, 1996. In an operation that probably reflected a blend of proficiency and good luck, Russian intelligence reportedly zeroed in on Dudayev's position and transmitted coordinates to VVS ground attack aircraft, which then fired radio-frequency homing missiles that targeted Dudayev while he was talking on a satellite field telephone.[100] According to press accounts, two missiles were electronically guided by signals bouncing between the portable phone's antenna and a relay satellite. Although the defense ministry has refused to confirm or deny this account of events, Dudayev's death was conceded by his key deputies shortly thereafter, making him possibly the first victim of Russian "information warfare."

A question yet to be answered is whether the VVS will view the continued stalemate in Chechnya as presenting a windfall opportunity to cycle aircrews in and out of the theater periodically for combat training under realistic conditions, as well as for operational test of new systems and mobility training in support of the new power-projection emphasis of Russian strategy. (As a straw in the wind on the second count, Russia has only two latest-generation Ka-50 Werewolf attack helicopters. Neither was sent to Chechnya for combat evaluation.) In all likelihood, a continued lack of funds will inhibit the fulfillment of any VVS interests along these lines.

Should the VVS continue to be tasked with an operational role in Chechnya, its involvement will almost surely assume new contours. In particular, it will face a more demanding task in keeping the surviving irregulars bottled up in their mountain hideouts. Problems of

---

[99]Ibid.

[100]See Richard Boudreaux, "Chechens Drop Russia Talks After Leader's Death," *Los Angeles Times*, April 25, 1996.

target location and identification will be severe, much as they have been for NATO air forces operating over Bosnia. The downside for the Russians is that they have far less operational and technical capability than NATO.

The extent to which the VVS was strapped in fulfilling its tasking in the relatively low-intensity war in Chechnya indicates that as long as it remains financially deprived, it will constitute, at best, only a regional air arm with little sustainability or capacity for high-technology combat. Considering that it faces no challenge that would justify a force structure more capable than a reconstituted variant of what it already possesses, some of the VVS's declared acquisition goals, notably a fifth-generation air superiority fighter (discussed in detail in the next chapter), will continue to exceed its grasp. The VVS will require more modest and tailored ambitions, at least for the near term, if it is to recover and flourish as a viable institution.

Chapter Nine

PROSPECTS FOR A FIFTH-GENERATION AIR SUPERIORITY FIGHTER

One of the most intriguing questions about the VVS's fitful adjustment to the post–cold war era concerns the near-term likelihood of its producing and deploying a new air superiority fighter to replace the current fourth-generation MiG-29 and Su-27. This question has an important policy implication for the United States as well. It was, after all, the strong possibility of such a development before the USSR's collapse that partly underlay the initial rationale for the USAF's F-22 fifth-generation Advanced Tactical Fighter (ATF).

Roughly a decade ago, the U.S. government predicted that two follow-on aircraft to the MiG-29 and Su-27, then generically labeled the Air Superiority Fighter (ASF) and the Defensive Counterair Fighter (CAF), would enter production in the mid-1990s. Following the 1988 Farnborough Air Show, however, Soviet aerospace officials led Westerners to believe that the only fighter prototypes they were working on were developmental variants of the MiG-29 and Su-27. This was later underscored by the Soviet Air Force's then–commander in chief, Colonel General Shaposhnikov, when he suggested that two new fighter types planned for deployment in the late 1990s would be step-upgrades of the MiG-29 and Su-27.[1] By 1990, the U.S. Defense Department's annual *Soviet Military Power* had scaled back earlier projections of a fifth-generation replacement for the MiG-29 and Su-27, noting merely that the possibility of a new

[1]See Nick Cook, "Soviets to Deploy Two New Fighters," *Jane's Defense Weekly*, July 27, 1991, p. 132.

Soviet fighter appearing sometime after the turn of the century "remains a concern."[2]

More recently, the U.S. Air Force has declared its expectation that a new Russian air superiority fighter will be test-flown by 1997 and will enter squadron service one to three years after the F-22's scheduled attainment of initial operational capability (IOC) in 2004.[3] It has further projected that Russia will field four wing-equivalents of this new fighter by 2020 and will produce additional numbers for sale abroad.[4]

In a highly publicized opposing argument in early 1994, the General Accounting Office (GAO) invoked the authority of Defense Intelligence Agency assessments to maintain that deployment of a Russian F-22 counterpart is *not* a realistic prospect for the near-term future. The GAO used that premise as the linchpin of a recommendation to Congress that development and production of the F-22 be deferred until at least the first decade of the next century.[5]

The USAF's projection of a possible initial test flight by a Russian next-generation fighter by 1997 is not unreasonable, given what we think we know about where Russia currently stands with respect to the development of such a fighter. Indeed, as the following sections will argue, it could occur even sooner, any day now for that matter, if repeated assurances by both General Deinekin and the Mikoyan Design Bureau can be believed. However, the projection of an IOC date before 2010 is based on the Herculean assumption that post-Soviet Russia can rebuild its shattered economy to the point where the needed funding to underwrite a full-scale advanced fighter production effort will become available in the time required. There are good independent reasons why the United States should remain com-

---

[2] *Soviet Military Power: 1990*, U.S. Government Printing Office, Washington, D.C., 1990, p. 80.

[3] See "USAF Promotes MFI to Defend F-22," *Flight International*, April 6-12, 1994, p. 5, and "USAF Prepares Rebuttal to GAO F-22 Report," *Inside the Air Force*, March 11, 1994, p. 6.

[4] See "USAF Expects Russia to Field Four Wings of Multirole Fighters by 2020," *Inside the Air Force*, April 1, 1994, p. 3.

[5] *Tactical Aircraft: F-15 Replacement Is Premature as Currently Planned*, Report to the Chairman, Legislation and National Security Subcommittee, Committee on Government Operations, U.S. House of Representatives, Washington, D.C., U.S. General Accounting Office, GAO/NSIAD-94-118, March 1994.

mitted to replacing the F-15 with a new air superiority fighter beginning in the first decade of the next century, *irrespective* of what Russia may be up to in advanced fighter development and planned deployment.[6] However, any projection of a fifth-generation Russian fighter achieving operationally meaningful strength in the foreseeable future would, in and of itself, constitute a weak foundation on which to base a case for continued F-22 development, as the following assessment will show.

RUSSIA'S INTEREST IN A FIFTH-GENERATION FIGHTER

If Soviet communism and the USSR had not disintegrated, there is no question that the USAF's F-22 would have prompted the development and deployment of a Soviet counterpart in due course. It is now widely recognized that both the formal issuance of military requirements for new Soviet combat aircraft and the actual design features of those aircraft were directly responsive to development initiatives in the West. On the first count, Russian engineers have repeatedly stated in conversations with Westerners that Soviet fighters were long developed and deployed in reaction to U.S. aircraft. One flatly remarked that the Su-24, Su-25, Su-27, and MiG-29 were conceived as Soviet "answers" to the F-111, A-9/A-10, F-15, and F-16. "Without the F-15," he said, "there would never have been a Su-27. Without the F-16, there would never have been a MiG-29."[7]

On the second count, the Soviet tendency was to emulate what had already been successfully demonstrated elsewhere. This inevitably resulted in a systemic lag in the appearance of new aircraft intended to offset their Western counterparts. A former Soviet engineer with over 18 years of helicopter design experience has written that new Soviet aircraft developments were "usually authorized, financed, and supported only after they have actually been realized in the West.

---

[6]For a thumbnail sketch of this case, see Benjamin S. Lambeth, "To Dominate the Skies: Why America Continues to Need the F-22," *Armed Forces Journal International*, November 1995, pp. 35–37.

[7]For more on this, see Benjamin S. Lambeth, *From Farnborough to Kubinka: An American MiG-29 Experience*, Santa Monica, California, RAND R-4000-RC, 1991, pp. 118–121.

The military first receives all new information on Western developments, and fear of dropping behind spurs them into swift and energetic action. New ideas are then supported and financed, and the R&D assumes national significance."[8]

Even with the Soviet Union gone and Russia now struggling to play a more cooperative role in world affairs, strong incentives remain among its defense elites to develop and produce a next-generation fighter. This interest has to do partly with Russia's natural competitive urge. Shortly before the USSR collapsed, the Soviet view was that, notwithstanding major changes for the better in Soviet-American *political* relations, U.S. combat aircraft would remain the principal rivals of Soviet aircraft, requiring that Soviet efforts to improve their aviation technology "should be comparable to corresponding steps by the Americans."[9] This view has persisted among Russia's aviation professionals in the post-Soviet era.

The incentive to pursue a new Russian fighter also has to do partly with Russia's sense that it lags behind the West in technology applications and has an obligation to catch up. The Mikoyan Design Bureau's chief, Rostislav Belyakov, admitted during an invitational visit to RAND in 1989 that Soviet avionics were inferior to those of the West in weight, size, and power consumption and that improved efficiency was a goal toward which redoubled efforts were required. He added that avionics integration was his biggest problem. Sukhoi test pilot Viktor Pugachev likewise has reported a higher Russian avionics fraction, conceding that in weight, if not in performance, Russian electronics remain inferior to Western electronics and that resolving this deficiency "will take time."[10] Following his appointment as Soviet defense minister in the wake of the abortive 1991 coup, Marshal Shaposhnikov concurred that the USSR was "behind."[11] He also stressed several times that Moscow could not

---

[8]Lev Chaiko, *Helicopter Construction in the USSR*, Falls Church, Virginia, Delphic Associates, Inc., 1985, pp. 70–71.

[9]Vladimir Ilin, "The Fighters of the 21st Century," *Krylia rodiny*, No. 4, April 1991, p. 20.

[10]Interview by Yu. Morozov, "This Is Our Life," *Morskoi sbornik*, No. 7, July 1992, pp. 43–48.

[11]Interview by B. Moseichuk, "We Are Shedding Excess Weight," *Argumenty i fakty*, No. 38, September 1991, p. 2.

accept settling for a "second best" position in relation to its new American "partner."[12]

Partly too the appeal of a new air superiority fighter has to do with Russia's determination to emerge from the wreckage of communism with more than just the appearances of a great power. Belyakov left little doubt about his own thoughts on this score when he declared at the 1991 Paris Air Show: "We know how the F-22 will perform. Therefore, we must also have a new fighter."[13] Mikoyan's now-retired chief test pilot Valery Menitskii has stated emphatically, even passionately, that if the United States needs the F-22, Russia has every need for an advanced fighter of comparable sophistication and performance.[14]

Finally, interest in developing a follow-on to the MiG-29 and Su-27 has to do with the opportunity a new fighter might offer Russia for remaining a credible competitor in global aerospace technology and thus attracting much-needed hard currency through foreign military sales. Throughout the Soviet period, new fighters would not typically find their way into the hands of Soviet allies and client states until well after they had been integrated into the VVS and VPVO inventories. Today, however, with the increased independence of Russia's aviation industry and its demonstrated willingness to sell virtually *anything* to foreign buyers ready to pay the price, such an aircraft could be marketed aggressively in the international arms forum.

All in all, Russian air power specialists have had enough to say about stealth applications over a long enough period for us to know that their interest is more than academic. Furthermore, there have been numerous pronouncements over the past four years from diverse Russians in a position to know stating that Russia's aircraft industry has progressed well down the road toward actually developing and producing a fifth-generation fighter prototype. With due allowance for the self-interest behind some of these pronouncements, it would be imprudent not to pay them serious attention and respect.

---

[12]Interview on Radio Rossiia, Moscow, September 18, 1991.

[13]Quoted in Michael D. Towle, "Stealth Jet Doesn't Overawe Soviet Expert," *Fort Worth Star-Telegram*, June 21, 1991.

[14]Conversation with the author in Moscow, September 5, 1993.

APPRECIATION OF THE TACTICAL ADVANTAGES OF STEALTH

Russian experts have followed the USAF's advanced tactical fighter activities closely from their inception. They have also given much thought to the role of stealth and its impact on the operational arena. Many indicators exist that the Russians well appreciate the importance of stealth in air combat. One commentary in 1991 correctly noted that with the advent of low-observable air superiority fighters, search and detection effectiveness will become the dominant factor in shaping the contours of an air battle.[15]

The author of this commentary, a VVS colonel with appropriate technical background, predicted that stealthy air superiority fighters armed with launch-and-leave missiles would put conventional fighters at a pronounced disadvantage in detection range, especially if the attackers entered the fight with their radars in standby and if situation awareness were provided by secure off-board C3 (command, control, and communications) such as that from an AWACS. Because of markedly dissimilar weapons engagement ranges (the author suggested 70 km for a stealthy attacker versus 15 km for a conventional defender), no defender would willingly press an attack to the merge—the point at which visual contact with the enemy is established and the engagement is joined—knowing that his destruction by enemy missiles could occur before he could establish a firing solution of his own. In a scenario in which *both* sides had stealthy fighters, the author suggested that aerial combat might revert to the within-visual-range tactics of the 1950s and 1960s, with front-quarter attacks effectively ruled out because of short detection ranges and with engagements devolving once again into stern conversions out of head-on passes at the merge.[16]

The author conceded that "no complete answer" is yet available to the question of how a conventional defender might cope effectively with a stealthy air-to-air threat. He suggested that one promising

---

[15]Colonel A. Krasnov, "Air Combat with 'Ghosts,'" *Aviatsiia i kosmonavtika*, No. 7, July 1991, pp. 4–5.

[16]This assumes, of course, that one or the other side would be foolhardy enough to insert expensive ATFs intentionally into multiparticipant maneuvering dogfights, with their attendant risk of higher loss rates on both sides because of unobserved shots.

area might include such options as jamming the enemy's missile radar and the clever use of deceptive tactics to degrade a technically superior attacker's situation awareness. He concluded, however, that the appearance of a next-generation fighter with a predicted radar cross-section as low as 1 percent of that of the F-15 or F-16 would render working out effective techniques of aerial combat "a task for tomorrow."

A RUSSIAN ADVANCED FIGHTER PROTOTYPE IN HAND?

There is no "smoking gun" evidence that Russia is at the point yet where it can claim to have a fully built F-22 equivalent waiting in the wings somewhere on a factory or hangar floor. There has been no shortage of authoritative comment, however, to the effect that some such Russian airframe exists, with the prospect of a public unveiling and first flight at any time.

In one of the first of such references, Mikoyan's Belyakov stated to a French reporter in 1991 that his design bureau was hard at work on two new fighters—a counterpart to the American ATF and an analogue of the French Rafale.[17] Belyakov said he was not confident that Mikoyan's experimental ATF prototype then in development would be the basis of the next advanced Soviet fighter. He did allow, however, that the effort had been given a project number indicating a formal development start in early 1988.[18]

During his short tenure as the VVS's commander in chief, then-Colonel General Shaposhnikov also asserted that the Soviet aviation industry was working on two new "stealth-optimized" fighters in response to VVS requirements, as well as on new air defense systems for negating stealth-class weapons. He said that any ultimate deployment of these systems would depend on whether the United States continued with the B-2 and the F-22, both of which he charac-

---

[17]The latter, it now seems apparent, was the MiG-29M.

[18]Sukhoi was also reportedly engaged in work on "fundamentally new types of fighters." However, it was said to be concentrating in the main on improving the Su-27. *Krylia rodiny*, No. 4, April 1991, p. 21.

terized as offensive weapons that had no place in the post–cold war era.[19]

A false rumor, evidently triggered by a comment by the then-president of the Russian Aviation Union, Vladimir Laptev, circulated in early 1992 that prototypes of Russia's next-generation counterpart to the U.S. ATF from both Mikoyan and Sukhoi were "under flight test." That report further claimed that despite a 50-percent reduction in Russia's military R&D budget since 1991, Sukhoi and Mikoyan had been directed to focus on core military programs, including new fighters.[20]

By early 1993, Belyakov volunteered that a Mikoyan fifth-generation fighter prototype was in hand but had not flown because of delays in engine development. Said Belyakov: "A completely new aircraft has been assembled. We had been planning its first flight in 1991, but we could not have done it. There is no money to continue the development of its engine. The airframe and the engine pioneer new technologies."[21]

Belyakov did not identify the aircraft, but conjecture was that he was referring to "Object 1.42," the reported company designator for the project. This report called the aircraft a Rafale look-alike and characterized it as the MFI, or *Mnogofunktsionalniy frontovoi istrebitel* ("Multirole Frontal Fighter") commissioned by the VVS.[22] Because of

---

[19]Interview by Nick Cook, "Soviet Air Chief Pushes for Quality," *Jane's Defense Weekly*, August 10, 1991, p. 220.

[20]Nick Cook, "Advanced Fighters Under Flight Test," *Jane's Defense Weekly*, March 7, 1992, p. 373. This report also referred to Sukhoi's proposed Su-37, which was a company idea for a light multirole fighter along the lines of the Swedish Gripen. By all indications, it is not being funded by the VVS and has never advanced beyond the paper stage.

[21]Quoted in "Lack of Funds Holds Up Mikoyan's F-22 Rival," *Flight International*, March 10–16, 1993, p. 5. A flight of the MFI in 1991, only a year after the first flights of the USAF F-22 and F-23, would have been unprecedented in Russian fighter development. Hitherto, the pattern of new Soviet fighter design typically featured a lag time of four to five years or more between the initial flights of the American and offsetting Russian aircraft.

[22]In seeming contradiction to these hints from Mikoyan, General Deinekin told German Luftwaffe sources in 1992 that a Russian ATF was *not* in development. See Charles Bickers, "Russians Trying to Veil New Fighter," *Jane's Defense Weekly*, August 22, 1992, p. 5.

problems with the new Al-41 engine being developed by Lyulka/Saturn, it was suggested that the first of two 1.42 prototypes believed to have been completed could fly with interim Al-35 engines currently being used in the Su-35. Later reports indicated that the new engines have been installed on the fighter and that they are full-scale development items mounting circular thrust-vectoring nozzles. According to Lyulka/Saturn president Viktor Chepkin, the A1-41 operates at a turbine-inlet temperature of 250 degrees higher than that attainable by the A1-35. Chepkin added that his company's product "is in no way inferior to the Pratt and Whitney F-119 engine powering the F-22."[23]

The aircraft itself has been variously described as having a blended body design similar to that of the F-22 and a canard/delta planform more like that of the Gripen or Eurofighter, in either case featuring twin engines, twin vertical stabilizers, and the use of shaping and radar-absorptive material to enhance its low observability. Mikoyan's deputy general designer, Anatoly Belosvet, has said that the airplane is in the 30-metric-ton class (about 66,000 lb), which would put it above the normal gross weight of an F-15C.[24] A French journal speculated that the aircraft is designed to carry the R-77, R-73, and R-37 air-to-air missiles internally, much in the manner of the F-22.[25]

Belyakov has lobbied for the 1.42 project personally on several occasions with President Yeltsin and has conceded that the defense ministry has not provided the needed funds to proceed with preparation for flight testing: "They have a requirement for this, but not the money yet." In the face of the budget crisis, he has argued for at least concept validation testing to maintain the design bureau's technology base and potential for developing a new fighter sometime in the first quarter of the 21st century.[26] Later, Belyakov claimed that development of Object 1.42 was continuing apace and that the first

[23]See "Lyulka/Saturn A1-41 Ready for Takeoff," *ConCISe*, August 31, 1995, p. 351.

[24]Cited in David Markov, "Russia's Hot New Fighters," *Air Force Magazine*, September 1993, p. 40.

[25]"Les Russes Modernisent et Visent l'Export," *Air et Cosmos*, No. 1467, April 11, 1994, pp. 33–34.

[26]Craig Covault, "Russia Debates Doctrine, Bomber, Fighter Decisions," *Aviation Week and Space Technology*, May 31, 1993, p. 23.

flight could take place before the end of September 1994.[27] At the Farnborough Air Show earlier that month, both he and General Deinekin reiterated during press conferences that the first flight of Mikoyan's MFI prototype was "imminent."

The chief of the VVS headquarters staff, Colonel General Anatoly Malyukov, underscored in early 1993 the VVS's interest in continuing with Object 1.42. Seemingly contradicting Belyakov's complaint that the VVS is not supporting the new fighter's development, Malyukov said: "We're not going to kill the program. We will try by all means to support this work, and we have some more freedom in the form of nonbudgetary funding, which has not been available up to now."[28]

Mikoyan was also said to have in development another aircraft called "Project 701," its response to a VPVO requirement for a long-range multirole interceptor (*Mnogofunktsionalniy dalniy perekhvatchik*) to replace the MiG-31. A later report from Mikoyan stated that Project 701 had been cancelled. According to a deputy chief designer, Vano Mikoyan, "there is no money for this project. It was simply too expensive."[29] The latter aircraft, with a cranked-delta wing and a projected Mach 2.2 cruise capability, may have been terminated as early as 1991.[30]

CONSTRAINTS ON THE NEAR-TERM EMERGENCE OF A NEW RUSSIAN FIGHTER

There is little denying that Russia has both the technical means and the incentive in principle to develop and deploy a fifth-generation air superiority fighter. Indeed, the Mikoyan Design Bureau very likely

---

[27]See Douglas Barrie, "Mikoyan Poised to Fly F-22 Rival," *Flight International*, July 13–19, 1994, p. 12, and Charles Bickers, "Multirole MiG Fighter to Fly This Summer," *Jane's Defense Weekly*, July 16, 1994, p. 1.

[28]Interview by Piotr Butowski, *Jane's Defense Weekly*, April 17, 1993, p. 15.

[29]Quoted in "Lack of Funds Kills Russian Interceptor Program," *Flight International*, November 17–23, 1993, p. 11.

[30]A Mikoyan drawing of the aircraft in an executive passenger configuration, called the MiG-701P (P for *passazhirskii*), was displayed at the 1993 Moscow Air Show. Mikoyan's marketing of the aircraft in this mode may be an indication of the state funding problems it has experienced. See Piotr Butowski, "Russians Planning New Long-Range Fighter," *Jane's Defense Weekly*, October 30, 1993, p. 5.

has a prototype successor to the MiG-29 in hand which it could fly at any time. Its leaders have long hinted at such a prospect in the most emphatic terms. For its part, the VVS has recently assigned the aircraft the designator MiG-37, in a departure from its long-standing practice of not assigning a new Mikoyan aircraft a "MiG" number until the aircraft is formally accepted by the VVS.[31] Accordingly, a rollout and maiden flight of such an aircraft should come as no surprise if and when it finally occurs. However, because of its economic crisis, Russia lives in a world of severe fiscal constraints that militate against the full-scale production and deployment of a next-generation fighter any time soon, even if Mikoyan were to test-fly a prototype successfully.

To begin with, Russia faces no air-to-air or other operational challenge today that would even remotely warrant the expenditure of large amounts of scarce funds on the VVS's stated need for a new air combat fighter. With the disappearance of the United States and NATO as military threats to Russia, any potential military problem that Russia may face in the near term will be regional in nature. For that reason, the main thrust of Russia's current defense planning is being directed toward power projection, not high-technology air warfare, as the VVS's recent combat experience in Chechnya has amply attested.

Second, as noted earlier, Russia is rapidly retiring much of its *existing* fighter force. General Deinekin said that the 13,000 pilots and 5000 aircraft inherited by Russia from the former USSR made for a "more than inadmissible armada of people and equipment," considering Russia's greatly diminished post–cold war operational needs. He said his first priority was to trim this force down, eliminate many of the "several dozen" different types of equipment in the existing inventory, and concentrate on creating a leaner force based on the most current bomber, fighter, and transport types now in operational service.

As for new acquisitions, General Deinekin said that appropriations from the Ministry of Finance were insufficient to refurbish the VVS, and that for the moment he was looking to procure new hardware

[31]See Simon Saradzhyan, "Russia Spending Scarce Funds on Futuristic Warplanes," *Radio Free Europe/Radio Liberty Report*, April 22, 1996.

only "in the minimum amounts that can assure the survival and profitability of aircraft plants."[32] First Deputy Defense Minister Kokoshin likewise confirmed that the defense budget authorized by the Ministry of Finance would permit little more than providing the armed forces with "at least a minimal amount of new equipment."[33] A year later, Deinekin reported that Russia's economic crisis had forced the VVS to abandon the development and production of a number of new equipment types.[34]

Competing Air Force Investment Priorities

On the books, the VVS has a stated requirement for a follow-on to the MiG-29 and Su-27. The first deputy commander in chief, Colonel General Viktor Kot, reported in January 1994 that the main efforts of VVS acquisition planning were being targeted, among other things, on the "top-priority development of fifth-generation aviation complexes."[35]

Acquiring a new air superiority fighter, however, is far from the VVS's most immediate concern. Even before the collapse of the USSR, General Malyukov indicated that the VVS would face a severe fiscal challenge in attempting to replace the MiG-29 and Su-27 with a new aircraft type. As for stealth applications in general, he said: "We are working on it, as are all nations that develop their own aircraft." However, he declared that the USSR was not developing any combat aircraft like the USAF's F-117 in which all other performance attributes would be subordinated to stealth. This echoed similar comments from Soviet industry representatives. More ominously for the near-term prospects for Object 1.42, Malyukov added that "we find ourselves in a difficult position in funding research and devel-

---

[32]Interview with Colonel General Petr S. Deinekin, "A Russia Without Wings Is Not Russia. It Does and Will Have Them," *Krasnaia zvezda*, August 15, 1992.

[33]Interview on Moscow television, September 17, 1992.

[34]Colonel General Petr S. Deinekin, "Always Ready for Action," *Krasniy voin*, August 7, 1993, p. 3.

[35]Interview with Colonel General Viktor Kot, "The Air Force Acquires a New Image," *Krasnaia zvezda*, January 27, 1994.

opment."[36] In light of that, he suggested that "it might be better to cut back on procurement and leave funding for R&D, even though this may mean stopping production factories and putting thousands of people out of work."[37]

An earlier Soviet discussion of ATF matters indicated that current fourth-generation fighters had limited growth potential and that any curtailment in the development of follow-on systems would occasion "a technological lag in one of the areas where we still hold world-class positions." That report frankly added, however, that serious financial difficulties had arisen in the defense sector, such that "preference has been given to the radical upgrading of existing aircraft rather than the creation of fundamentally new designs."[38]

General Deinekin has declared that the main goal of VVS force modernization to the year 2000 is the creation of a mobility capability to support Russia's peacekeeping needs around its conflicted periphery, notably in Transcaucasia and Central Asia. He has added that acquisition of new transport aircraft is the VVS's top procurement need.[39] This is not surprising, considering that power projection has become the dominant focus of Russia's new military doctrine.

General Deinekin has stated that the VVS's cessation of further MiG-29 and Su-27 procurement was not occasioned by budget restrictions, but by a determination that the number already in hand was sufficient to meet Russia's current operational needs. As for the future, he said that the VVS will seek to acquire next-generation aircraft and weapons, but only "as Russia's economic situation stabilizes."[40] In the meantime, it has had to defer even the development and production of several variants of the Su-27, which have been

---

[36]Interview by Dmitri Grinyuk and Piotr Butowski, "An Unusual Conversation at the Main Staff," *Krylia rodiny*, No. 11, November 1991.

[37]My RAND colleague Abraham Becker has pointed out that any such reallocation of resources *could* mean more money for long-lead ATF development.

[38]*Krylia rodiny*, No. 4, April 1991, p. 21.

[39]Interview with Colonel General Deinekin by Charles Bickers, *Jane's Defense Weekly*, May 7, 1994, p. 32.

[40]Interview with Colonel General Deinekin by Gennady Lisankov, "Aircraft First," *Rossiiskaia gazeta*, May 5, 1994.

designated by the VVS as the intended mainstays of Russia's fighter inventory for at least the remainder of the 20th century.

A poignant reflection of the dire straits in which Russia's military aviation industry now finds itself was the proposal by the head of the VVS's Central Research Institute, Major General Vasily Aleksandrov, at an international air power conference in Great Britain that the decline in Russia's defense funding had generated sufficiently "troublesome financial barriers" to argue for Russian "effort integration" with potential foreign partners. General Aleksandrov suggested that Russia be considered for inclusion in a multinational requirements validation for an advanced "Euro-Russian Fighter Aircraft" (ERFA) as "an important element for peacekeeping forces of the future."[41]

Competing VVS Development Options

Object 1.42 is also not the only item on the VVS's current force development plate. Russia has other programs in train in the air-to-air mission area offering potential returns at a fraction of the cost of a new fighter. These include, among other things, the Su-35, Fazotron's *Zhuk* AI radar, and an active radar missile comparable to the American AIM-120 AMRAAM.

On the first count, what the U.S. Defense Department had for several years postulated as an eventual new Soviet Air Superiority Fighter began by early 1991 to take on the familiar shape of an advanced variant of the Su-27, with improved maneuverability and nose authority at high angles of attack, more fuel-efficient engines, greater thrust, and a digital flight control system. This major improvement over the Su-27, as noted earlier, is now called the Su-35. It falls far short of offering anything like the low observability and supercruise performance promised by the USAF's F-22. However, it portends at least a credible match for the F-15.[42]

---

[41]Major General Vasily E. Aleksandrov, "The Prospects for Air Superiority Fighter Development," paper presented at the First International Air Power Conference sponsored by British Aerospace, London, England, February 11–12, 1993.

[42]"Flanker Follow-on," *Aerospace Daily*, November 25, 1991.

There is also an effort in train to upgrade the MiG-29 with a new radar, called *Zhuk* ("Beetle"), now being developed by Fazotron. It features a slotted array antenna and new weapons integration modes capable of accommodating Russia's AMRAAM counterpart. Deputy chief designer Valery Antonov stated that the radar was developed in response to a joint request from Mikoyan and the VVS. It offers a track-while-scan capability and can engage up to four targets simultaneously if the weapon has an active seeker. The previous MiG-29 radar had only a single-target track capability.[43]

Finally, Russia has developed and successfully tested an all-aspect, launch-and-leave missile with capabilities comparable to those of the U.S. AMRAAM. A mockup of this missile, the Vympel R-77 (AA-X-12 Adder), was displayed at the 1992 and 1993 Moscow Air Shows. According to a Russian technical publication, development of the weapon has been under way for ten years.[44] It features an active radar seeker, low-aspect aerodynamic surfaces, and four latticed control surfaces mounted on the tail section to increase the missile's agility and reduce its radar cross-section. The missile is capable of being fired at targets up to 90 degrees off boresight, and plans are reportedly in hand to develop a variant capable of engaging AWACS-type targets at ranges of up to 150 km. Tests have been completed and the missile is awaiting a production decision. Deliveries for export have also been proposed.

The seeming lack of total commitment on the part of the VVS's leadership to Mikoyan's putative MiG-37 project in the face of these more affordable options was reflected in what came across as, at best, a lukewarm endorsement of the program by General Malyukov. With respect to Mikoyan's financial needs to keep the program alive, he said: "We can help them with cash injections, but it is hard to look ahead." He frankly added that the VVS was torn between continuing with the follow-on fighter and concentrating its meager resources on more immediate force development priorities: "We might save this program through a big investment, but we are in a complicated posi-

[43]"Mikoyan Tests Radar Upgrade for MiG-29," *Flight International*, August 26–September 1, 1992, p. 17.

[44]"The Mosaeroshow-92 Russian Aerospace Exhibition," *Byulleten inostrannoi nauchnoi i tekhnicheskoi informatsii*, No. 41, October 13, 1992.

tion, because the Su-27 is in production and because proposed modifications are wide-ranging and *in principle will satisfy our requirements*" [emphasis added].[45]

General Malyukov further stated that because of cash shortages, the VVS's first priority in fighter force enhancement would be Su-27 modifications, including the Su-35 air superiority fighter and Su-32 two-seat all-weather strike aircraft, as well as a reconnaissance variant and a much-needed ECM/EW version.[46] He also conceded that the VVS's most pressing acquisition need was in the transport area, considering Russia's new power-projection emphasis and the fact that more than half the VVS's Il-76s were lost to Ukraine as a result of the USSR's breakup. The vice president of Aviaprom, Viktor Laptev, has concurred that the current emphasis in Russian weapons development policy is on upgrading existing platforms. This casts yet another shadow on the outlook for the would-be MiG-37.[47]

Competing Defense Ministry Priorities

Whatever the stated requirements of the VVS may be, it does not command the inside track when it comes to making core decisions on defense resource apportionment. Russian aircraft designers have been among the first to acknowledge the old saw that he who pays the piper calls the tune. And that, in this case, is the Ministry of Defense. The VVS will never get a new air superiority fighter without the defense ministry's support. And by all indications, the Ministry of Defense has more urgent priorities.

First Deputy Defense Minister Kokoshin declared that the defense ministry's intended strategy for stemming further dissipation of industry resources was to concentrate on tried and proven equipment and to forgo investment in programs that would not attain their projected specifications for years to come. At the same time, he said that

---

[45]Interview with Colonel General Malyukov, *Jane's Defense Weekly*, April 17, 1993, p. 15.

[46]Specifications and performance details for the Su-35 are presented in Andre Brand, "Le Sukhoi Su-35, un 'Vrai-Faux' Su-27!" *Air et Cosmos*, No. 1470, May 2, 1994, pp. 32–34.

[47]Quoted in "Lack of Funds Kills Russian Interceptor Program," *Flight International*, November 17–23, 1993, p. 11.

the ministry wished to accelerate work on pursuing new technological breakthroughs as a hedge against mortgaging Russia's defense future. He said nothing, however, about going from concept development to production.[48]

Whatever may be in the development pipeline, Kokoshin has stated that the defense ministry has already settled on its acquisition plans to the year 2000.[49] He has further indicated that decisions have been reached with respect to incremental growth in expenditures for weapons procurement, with two principal areas targeted for emphasis in the near term: first, modernization of systems that have shown the greatest performance, value, and growth potential; and, second, methodically laying down an R&D base that would enable "a certain qualitative leap and expanded series production of the most modern equipment *at a time when we are a little richer*" [emphasis added].[50] This suggests that although the defense ministry will try its best to keep a modicum of sustainer funding channeled into the MiG-37 program, any full-scale engineering and manufacturing development of the airplane will be subordinated, at least in the near term, to concentrating on improvement of the Su-27 and other fighters already in operational service.

Kokoshin admitted that Mikoyan "is developing a future fighter" and that the defense ministry would like "to order more promising machines," rather than waste any further outlays on additional MiG-29s, which already exist in adequate numbers. But he noted at the same time, by way of example, that it would be eight times less costly to upgrade the MiG-31 than to develop and produce a new long-range interceptor, and that an improved MiG-31M would have "almost the same tactical-technical specifications." He added that Russia's defense industry was "in a very serious position" with respect to the danger of falling behind. He called the current crisis "one of the most dramatic moments in all the many centuries of Russian history," with the pendulum having swung "from surpluses directly to

---

[48]Cited in Yury Mamchur, "If We Preserve the Defense Complex, We Will Preserve Russia," *Krasnaia zvezda*, October 29, 1992.

[49]Cited in a Radio Rossiia newscast, March 5, 1993.

[50]Interview on Moscow television, April 17, 1993.

the other extreme." Kokoshin described the decline in procurement which began in 1992 as "catastrophic."

A telling effect of the procurement crisis is that the VVS and VPVO together acquired only 17 new combat aircraft in 1993.[51] That number has gone downhill steadily in the years since, to the point where the 1996 budget provides for the purchase of no new combat aircraft. Given the bleak outlook for financing new systems and the problems the VVS faces in funding even the improvement of existing types, any next-generation Russian air superiority fighter fielded in unit strength may be a pipe dream until well into the 21st century. Only a few weeks before General Kot stated than such a fighter was a VVS "top priority," First Deputy Defense Minister Kokoshin declared that upcoming R&D and procurement for all services would focus mainly on reconnaissance, command and control, supply to mobile forces, and precision munitions.[52]

The Absolute Funding Shortage

Even if the Russians had the technology in hand and the VVS and defense ministry could both plead an airtight case for a next-generation fighter, the question would remain: How will they pay for it?

During a visit to RAND in 1989, Mikoyan's chief, Rostislav Belyakov, pointed out that a combination of declining defense resources and the recently declared Soviet defensive military doctrine could lead to pressures to curtail, or even forgo altogether, the deployment of successor-generation combat aircraft of all types. In the face of this, said Belyakov, it would remain crucial to continue pursuing new design concepts aggressively, since that is where technological progress ultimately comes from. Belyakov added, however, that because of the mounting cost of new weapons systems, it would become increasingly difficult to proceed from technology demonstrators to series production without an all but ironclad military justification. That was *before* the collapse of the USSR. Belyakov has since reported that his firm has been getting by on a

[51]David Hearst, "Military Cuts Put Yeltsin Under Threat," *The Guardian*, June 15, 1994.

[52]ITAR-TASS World Service, December 24, 1993.

"starvation diet" and that he is obliged to return 60 percent of his meager annual R&D allocations to the state budget to pay taxes.

On the eve of the USSR's collapse, before he was appointed First Deputy Minister of Defense of the Russian Federation, Andrei Kokoshin predicted that the defense industry would receive virtually no production orders for new equipment in 1992, since the entire budget would be needed to feed, clothe, and house Russian servicemen and their families.[53] This trend toward a predominantly social-welfare orientation in the defense budget was confirmed by the chief of the Central Finance Directorate of the CIS armed forces, Lieutenant General Vasily Vorobyev, in February 1992. General Vorobyev reported that about 70 percent of the total defense expenditure approved by the Russian Federation Supreme Soviet in the first quarter of 1992 would go to the welfare needs of servicemen and their families.[54]

The general director of the aviation industry department of the Ministry of Industry, Anatoly Bratukhin, told reporters that the volume of military production orders was down by a factor of five in 1992 from 1991.[55] Of R7 billion earmarked for procurement (down from R32.6 billion in 1991), only R0.8 billion had been released during the first quarter. The situation for R&D was only slightly better. Only R0.8 billion was allocated the first quarter of 1992, down from an annual amount of R15.3 billion in 1989.[56] According to then–Prime Minister Yegor Gaidar, half the procurement funds spent in 1992 went directly to compensate producers for cancelled orders and to keep them temporarily out of bankruptcy.[57]

---

[53]John Lloyd, "Sharp Cut in Soviet Defense Orders," *Financial Times* (London), October 28, 1991.

[54]Interview with Lieutenant General Vasily Vorobyev, "The Priority Is for Social Programs," *Syn otechestva*, February 14, 1992.

[55]Cited in Yury Kovalenko, "The French and Russians Will Build a Helicopter," *Izvestiia*, June 12, 1992.

[56]It is hard to compare these two numbers, however, because of the strong inflation the USSR experienced in 1991 and President Yeltsin's lifting of price controls in January 1992. I am indebted to Abraham Becker of RAND for calling my attention to this.

[57]Brigitte Sauerwein, "Defense Conversion: Russia's 'Strategic Imperative,'" *International Defense Review*, No. 6, 1992, p. 734.

These cuts were so severe that Sukhoi was reported to be in danger of losing both the naval and the all-weather ground-attack variants of its Su-27. A Sukhoi official reported that production of the Su-27K carrier version had been halted, adding that this was *not* the result of a policy decision but simply a consequence of insufficient funds. The Su-27IB development (now called the Su-32) was also operating on a shoestring budget, with only two prototypes built and a decision on series production deferred for the indefinite future, notwithstanding the fact that the Su-27 had been chosen by the VVS over the MiG-29 as the airframe of choice for near-term Russian fighter force modernization.[58]

In a grudging eleventh-hour acknowledgment that defense was being underfunded, the Ministry of Finance, under duress, granted the defense ministry a 10-percent increase in its 1993 procurement allocation. Prime Minister Gaidar conceded that earlier cuts had been "premature."[59] This scarcely vitiated, however, reported hard times at the Su-27 production facility in Komsomolsk-na-Amure. The factory director complained that from a steady production rate, he was suddenly informed in February 1992 that all state orders had been cancelled and that his products "were not needed by anyone." This meant that his only source of income was a percentage of the hard-currency proceeds from a small Chinese order and that the latter would only cover his expected costs through the following October.[60]

The factory director further reported that aside from a handful of carrier variants, not a single new Su-27 had been delivered to operational units in 1993. He said that a number of previously ordered Su-27s stood unpaid-for and unclaimed on his flight line and that his

---

[58]See Douglas Barrie, "Cuts Put Flanker's Future in Jeopardy," *Flight International*, August 26–September 1, 1992, p. 15.

[59]Cited in Valery Begishev, "The Defense Industry Is Gaining the Upper Hand," *Lesnaia gazeta*, September 19, 1992.

[60]Interview with Anatoly M. Petrovich, general director of the Komsomolsk-na-Amure Aviation Production Association, "It Isn't Conversion That Causes Unemployment," *Krasnaia zvezda*, September 30, 1992.

plant had been notified by the defense ministry that there was no money allocated for a single Su-27 modification.[61]

By late 1992, spending on R&D and weapons production was down 68 percent from 1991. Even in constant prices, that would have been a blow to Russia's defense industry by any measure. In the event the comparison was *not* in constant prices, the effect would most surely have warranted Kokoshin's characterization of the cuts as "catastrophic." Either way, they necessitated the closure of 21 defense industry concerns, with another 130 expected to follow shortly. More than 400 firms had scaled back their activities to a 3–4 day work week, and engineering staffs had been cut by 15–20 percent. This was a natural continuation of a trend begun as early as 1988, when defense industry conversion started and the new defensive military doctrine was announced. That year, production contracts to industry fell 24 percent from the previous year.[62]

In 1992, the Ministry of Finance failed even to earmark the requisite funds to cover fuel quotas allocated to the VVS by the Ministry of the Economy. As a result, continuation training in line units dropped to crisis levels. The head of the defense ministry's finance directorate said that production and delivery of armaments were also "sharply cut," since "very insignificant appropriations" were allocated to the ministry to pay for them.[63]

In response to a question about how the VVS intended to deal with its situation in which R&D funding had been essentially frozen, General Deinekin confirmed that he had been forced to abandon a series of projects that would have assured a competitive position for Russia. He said: "There are no resources for basic research, development, or even the maintenance of a distinctive experimental base." He added that without such investment, there will be "no future for the country's aviation—either military or civil" and warned

---

[61]Interview with Petrovich by Colonel A. Andryushkov, "We Make the Best Interceptor in the World," *Krasnaia zvezda*, January 13, 1993.

[62]Cited in Interfax, Moscow, October 26, 1992.

[63]Interview with Lieutenant General Vasily Vorobyev, "The Military Budget: Myths and Reality," *Krasnaia zvezda*, February 10, 1993.

that "only the adoption of urgent measures at the very highest level will enable Russia to remain an independent aviation state."[64]

No doubt some of this dire language was intended for lobbying purposes, which would have rendered it an overstatement of a bad enough situation. There is no question, however, of the VVS's sense of beleaguerment as a result of its continuing funding crisis. During an interview in early 1994, General Deinekin admitted that "we are doing our best" with Mikoyan's new fighter project, but added that the lack of adequate financing was a major hindrance. For that reason, he concluded, it would be "not soon" when the most demanding new projects came to fruition.[65] That prognosis can only have been reinforced by the steady diversion of scarce funds from more needy defense accounts that has since been occasioned by Russia's continuing military misadventure in Chechnya.

The most recent reports leave a heavy question mark over the status and future of the 1.42 project. During the final weeks that preceded the August 1995 Moscow Air Show, expectations were high that Mikoyan's new fighter would be rolled out and put prominently on static display. Those expectations, repeatedly stoked by hints dropped by Mikoyan's leadership, were dashed at the eleventh hour by a reported decree from defense minister Grachev that it was "too early" to unveil the new product. By the end of the year, there were reports that the fighter would make its maiden flight "early" in 1996.[66] The persistence of such rumors about the aircraft's existence and imminent flight was enough to prompt unnamed American officials to state that the new fighter would have sixteen control surfaces driven by a fly-by-wire system and would embody limited stealth characteristics, although without any capacity for internal missile carriage.[67]

---

[64]Interview with Colonel General Petr S. Deinekin by I. Chernyak, "The Man Sitting in Moscow Is Going Into a Spin Again," *Komsomolskaia pravda*, March 17, 1993.

[65]Bickers interview, *Jane's Defense Weekly*, May 7, 1994, p. 32.

[66]Nikolai Novichkov, "The Russian MiGs Set for Flight Test," *Aviation Week and Space Technology*, January 1, 1996, p. 21.

[67]"U.S. View of MiG-1.42," *Aviation Week and Space Technology*, January 15, 1996, p. 19. An earlier report, citing as evidence the forward-folding control surfaces of the new Vympel R-77 air-to-air missile, maintained that these reflected an intention for internal carriage and that the missile was expressly developed for Object 1.42, in

The longer the rumored new fighter remains masked behind this "any day now" posturing, however, the more likely it is that the program's financial troubles are grave enough to cast doubt on whether it will ever produce a fifth-generation successor to the MiG-29 in squadron strength.[68] Even during less troubled times, Mikoyan's chief designer Belyakov conceded that because of the aircraft's size and cost, it would constitute no more than 20–30 percent of the VVS's fighter force. More recently, his principal deputy, Anatoly Belosvet, has urged a major review of current Russian combat aircraft programs, an action that could force the 1.42 program to die on the vine. Should the VVS eventually settle on an upgraded Su-35 with an AMRAAM-class missile to replace the Su-27, it would effectively skip a generation of fighter development as a result.[69] Belosvet has admitted that the VVS cannot afford to buy the MiG-37 in enough numbers to be operationally credible, and that the program may accordingly become a technology demonstrator for a more affordable fifth-generation fighter.[70] At the 1996 Berlin Air Show, he gamely insisted that the 1.42 prototype could still fly "in a few months" if the needed funding comes through.[71] He has conceded elsewhere, however, that the MiG-29M and MiG-31M may be his company's last new fighter designs to fly in this century.

---

keeping with a long-standing Soviet practice of designing a new air-to-air missile for each new fighter. See "Waiting for Russia's Stealth Fighter," *International Defense Review*, June 1995, p. 42.

[68]Such posturing also plays nicely into the hands of Russian cynics who are wont to maintain that the unofficial state religion is "Skoro Buddhism," a derivation from the Russian *skoro budet* (loosely translated as "any day now").

[69]Douglas Barrie, "Cold Comfort," *Flight International*, May 31–June 6, 1995, p. 26.

[70]Douglas Barrie and Alexander Velovich, "Russia Faces Procurement Crisis as Air Force Budget Is Slashed," *Flight International*, January 31–February 5, 1996, p. 22.

[71]"Special Report," *ConCISe*, May 18, 1996, p. 476.

Chapter Ten

THE LONG ROAD TO RECOVERY

In light of the daunting problems outlined in this study, coupled with increasing doubts about the near-term prospects for political and economic reform in Russia, one can fairly ask whether the sun is rising or setting on General Deinekin's VVS. To this question, the VVS chief would almost surely answer with cautious optimism born of conviction. He has repeatedly declared that the VVS has the needed talent, and he has expressed an appreciation of its past failings under Soviet rule, a vision of what needs to be done to correct them, and an abiding determination that military aviation in Russia will eventually recover to full health. The hard reality, of course, is that the main factors that will determine the ultimate course and outcome of the VVS's struggle for resurrection lie largely beyond General Deinekin's control. At bottom, the fate of Russia's air power, like that of the military establishment as a whole, is inseparably tied up with the fate of post-Soviet Russia itself.

Three years ago, the chief of the VVS's headquarters staff, Colonel General Malyukov, stated his belief that the worst of the crisis had passed and that the VVS had successfully weathered the initial maelstrom of post-Soviet change. Outmoded or counterproductive practices, he declared, were being done away with, and new relationships were forming within the VVS and between the VVS and other sectors of the defense community.

That turbulent passage was no mean accomplishment in General Malyukov's view. After the union's collapse and the loss of many VVS equities to the breakaway republics, "it was very difficult," he said, "for us to keep the rest of the pieces of our air force operating. Now we know, more or less, on what we can depend; which factories, which supply bases, which units. Basically the work in these areas is already done, and our thoughts are turned to building a powerful air force." Most of the problems now confronting the VVS, added Malyukov, "are long term, relating to combat readiness and air combat capability."[1]

Yet despite that optimistic view, the VVS's funding predicament remains both acute and doubly compounded by Yeltsin's continued inability or unwillingness to extricate Russia from the quagmire of Chechnya. Force modernization has all but ground to a halt. Even R&D for the improvement of existing systems is dead in the water. The VVS is barely managing even to meet the payroll needs of its officers. Fuel supplies are only adequate to enable a small percentage of line pilots to remain on operational flight status. Even these, at 25–30 hours a year, are at best only able to maintain landing currency and proficiency at basic aircraft handling in clear weather. Operationally meaningful air combat training has become a thing of the past in most fighter units. As General Deinekin recently noted, VTA pilots are still getting a passable amount of annual flight time because of the diverse mobility demands placed on them, notably by the still-festering crisis in Chechnya. "As for fighter pilots," he said, "they do not do much flying."[2] The situation has become an increasing threat to flight safety, and the VVS's accident rate has risen to bear it out.

On the brighter side, the VVS's withdrawal from Eastern Europe and the former Soviet republics, begun before the USSR's collapse in 1991, is now complete. That phased withdrawal placed a heavy burden on the VVS's annual budget and came at the expense of needed training and other operational support accounts. Assuming that future VVS budget allocations do not suffer a precipitous decline from the 1995 level, this may open up needed funds for a gradual increase in support for training and readiness.

---

[1]Quoted in Piotr Butowski, "Flying in the Face of Adversity," *Jane's Defense Weekly*, April 17, 1993, p. 15.

[2]Interview by Vladislav Listyev on Ostankino television First Channel, December 14, 1994.

Even in the best case, however, Colonel General Mikhail Sorokha, a deputy commander on the headquarters staff, has indicated that the VVS has no realistic chance of offsetting its recent losses in strength by "a growth in quality of armament and equipment within the next few years." More than half its inventory consists of second- and third-generation aircraft slated to be retired by the years 2000–2005. When that happens, only 20 percent of the current force will remain operational. The VVS's stated annual need is for 250–300 new aircraft to meet the demands of the defense ministry's declared power projection and mobility doctrine. The 1994 defense budget provided, however, for only 32 for all services. For the first time in the history of Russian military aviation, the 1995 budget provided for *no* new aircraft purchases. Seventy percent of the VVS's airfields cannot handle night or adverse weather flights. Of 209 airfields altogether, 133 are in need of major repair and modernization, for which there is no money. Because production of the workhorse Il-76 jet transport has ended, VTA's numerical strength will begin to decline sharply in 1997 when the first of these aircraft are retired.[3]

According to Moscow defense correspondent Pavel Felgengauer, the war in Chechnya proved that Russia's weapons are "outdated and ill-suited even for a local war with a second-rate enemy," and that the Russian military needs "up-to-date communications facilities, combat helicopters, and airplanes capable of operating at night, in bad weather, and so forth."[4] That statement is true enough as far as it goes, but it does not capture the main priorities, which have more to do with first making the VVS's *existing* hardware work.

It remains too early to say what the future holds beyond the broadest of generalizations. With respect to force modernization, General Deinekin and other VVS leaders have openly stated their near-term intentions and goals through the year 2000. These goals are not unreasonable for the sort of air force that would seem appropriate for post-Soviet Russia, given its likely operational challenges in the immediate years ahead. Yet because of the continuing budget crisis,

---

[3]Andrei Baranovskii, "The Russian Air Force Has Nothing to Fly," *Segodnya*, March 30, 1995.

[4]Pavel Felgengauer, "No Professional Military Can Be Foreseen in Russia So Far," *Segodnya*, May 4, 1995.

it is hard to imagine how the VVS can take more than the first steps in this direction at a time when its procurement power is all but nil and it is having trouble merely providing its pilots with enough monthly flying time to keep them free from accidents.

As for doctrine and concepts, the VVS has discarded its canonical "Warsaw Pact Air Operation Plan," if indeed it ever paid more than lip service to that plan in its routine training, and now confronts a need to develop new strategies consistent with the emerging mission requirements of post-Soviet Russia. Russia, however, has yet to develop a coherent and fully articulated foreign policy, or even, for that matter, a considered set of national interests upon which such a policy might be based. Accordingly, its much-vaunted "new military doctrine," published in late 1993, remains little more than a statement of broad principles for an ideal world. In the absence of a clear threat or a readily definable operational challenge, any attempt to produce a more detailed repertoire for Russian air power would be putting the cart before the horse.

There is no clearer testament to the acuteness of the many problems the VVS faces today than its continued inability to do much beyond intellectualize over the implications of the 1991 Persian Gulf War. As the fall of Soviet communism neared, the VVS had a ringside seat from which to observe the allied coalition's successful air campaign against Iraq. That campaign opened the eyes of Soviet airmen not just to what Western aviation could accomplish, but to what air power in general (including Russian air power) could do if properly equipped, configured, and applied. Unfortunately for the VVS, however, that realization dawned precisely as its own operational and institutional moorings had become loosened by the winds of international and domestic change.

Not long after the Gulf war ended, the head of the Soviet Air Force's Combat Training Directorate, Colonel General Borsuk, suggested that the Desert Storm experience was forcing the VVS "to consider opportunities for elevating the role of air forces in contemporary warfare, operational employment, and training, and to make appropriate corrections in all those areas."[5] The deputy chief of the

---

[5]Interview with Lieutenant General A. Borsuk, "Corrections Are Needed," *Aviatsiia i kosmonavtika*, No. 7, July 1991, pp. 2–3.

Operational Art Department at the General Staff Academy similarly pointed out that "the lessons of Desert Storm are of practical interest to us," adding that "it would simply be shortsighted to ignore them in creating the Russian armed forces." He stressed the pivotal role played by coalition air power in demonstrating, for the first time, that strategic as well as tactical goals can be attained directly through the use of precision firepower. He further noted how the VVS's recently concluded *Voskhod '93* exercise had indicated Russia's potential for conducting standoff attacks at great distances from home base.[6] So far, however, the VVS has been unable to go much beyond lip service in assimilating the most obvious teachings of the Desert Storm experience.

Nevertheless, the powerful role model provided for Russian air tacticians by the allied coalition's performance in Desert Storm, coupled with the subsequent lifting of many of the former inhibitions that blocked any serious effort at tactical reform in the VVS, makes it fair to speculate that at least some of the impending changes in Russian operational practice, once they take root, will show a heightened Western orientation. The air-to-air arena warrants special attention in this regard. Because improvement in air combat prowess is essentially cost-free, in that it turns largely on changed procedures rather than new equipment, the VVS is now positioned to begin applying whatever inclinations its best tacticians may long have harbored toward new directions in air-to-air training.

One constraint here, probably a preclusive one for the near term, concerns the extent to which even seemingly "low-cost" changes in tactical training may be preempted by a diversion of already scarce operations and maintenance funds toward providing housing and other needed quality-of-life improvements for officers and their families. Another constraint concerns where the VVS's Combat Training Directorate will find a suitable home-grown experience pool from which to develop and pass along to Russian pilots a fun-

---

[6]Major General Yevgeny Korotchenko, "We Need a 21st Century Military," *Krasnaia zvezda*, June 9, 1993. Another general expressed this point even more baldly: "Is there a measure of effectiveness of the military which we dream about? Yes, there is. The Gulf war. In 40 days, Western professionals routed an Iraqi army built according to the Soviet model, armed with our weapons, and trained by us." Major General Vladimir Dudnik, "Military Reform: Eight Years of Marking Time," *Moskovskiye novosti*, No. 37, September 13, 1992, p. 7.

damentally new air combat repertoire. As the USAF's hard-earned experience over the past two decades since the first days of Red Flag amply attests, it takes far more to develop and acquire such a repertoire than merely reading the right books.

Exactly how the VVS will respond to its unprecedented opportunity for change remains hard to say. Much will depend, in the near term at least, on the extent to which Russia's weakened economy will permit the channeling of enough funds into the VVS's operations and maintenance accounts to underwrite a training program commensurate with the VVS's apparent new latitude for improvisation. Farther down the road, much will also hinge on the extent to which the dismantling of the old Soviet order will yield a permanent change in the daily pattern of organizational life for Russian pilots and commanders.[7]

Yet despite the many difficulties outlined above, the VVS's predicament is far from hopeless. Perhaps most important, it has been granted an end to political controls, increased freedom of expression, genuine encouragement from higher headquarters for the exercise of initiative and independent judgment from below, and an easing or elimination of the most odious former Soviet operating rules and restrictions. All of this has been geared toward enhancing opportunities for talented pilots and commanders to achieve their fullest potential. The system remains slow to change, and old habits linger on. But at least the door to reform is now open. Toward that end, General Deinekin has acknowledged the value of going to school on the West's experience: "There is something to adopt from abroad. We must train our pilots to world standards."[8] He has further noted that his many trips to visit foreign air forces have offered him useful

---

[7]Quite apart from the continued drain of its scarce resources down the sinkhole of Chechnya, yet another barrier to any early VVS recovery could be what appears to be a massive Russian construction effort in the southern Ural mountains aimed at building an underground military complex whose roots go back to the Brezhnev era and whose intent remains unclear to U.S. officials. With a defense budget of only some $11–14 billion, Russia will scarcely be able to fund any serious program of military reform if its leaders insist on continuing with atavistic pursuits. For more on this activity, see Michael R. Gordon, "Despite Cold War's End, Russia Keeps Building a Secret Complex," *New York Times*, April 16, 1996.

[8]Interview with Colonel General Petr Deinekin, "Who Will Take Up the Sword?" *Krylia rodiny*, March 1993.

food for thought: "Much of what I got to see and learn, I feel, would be expedient and possible to adopt into the combat training and everyday activity of the Russian Air Forces even today."[9]

Nevertheless, lean years lie ahead for military aviation in Russia. Much will depend on the still-uncertain prospects for economic stabilization and reform throughout the country as a whole. General Deinekin has indicated that the VVS, to its credit, retained a positive and solutions-oriented attitude even during the darkest years of stagnation under Brezhnev's rule. In acknowledging that, he highlighted what may well be his air force's greatest strength. His core dilemma is that just as the VVS has come within reach of an opportunity to institute sweeping changes in its repertoire, it has been denied the financial means to take anything more than the first halting steps.

Whatever path General Deinekin and his deputies ultimately adopt for selective belt-tightening and otherwise making the most of a grim situation, Russia's air force is well into its post-Soviet restructuring and has shown appreciable progress in adjusting to its new realities. Its leaders have candidly admitted their problems and have indicated what they believe needs to be done to start fixing them. This has removed a major obstacle from the road to recovery. It has also set the stage for a time of creative ferment, once Russia emerges from its current crisis with a measure of fiscal solvency.

---

[9]Interview with Colonel General Petr S. Deinekin, "A Time for Decisive Actions," *Aviatsiia i kosmonavtika*, No. 1, January 1993, p. 4.